Division

4TH – 8TH

Designs in Math

WRITTEN & ILLUSTRATED BY

RANDY L. WOMACK, M.ED.

LEARNING DISABILITIES & ART

PUBLISHED BY

G.E.C. PUBLICATIONS golden educational center

"LEADING THE WAY IN CREATIVE EDUCATIONAL MATERIALS" ™

857 LAKE BLVD. ❖ REDDING, CALIFORNIA 96003

Teacher – Parent Instructions For *Designs in Math*

This book has been written to aid the classroom teacher or parent in activities that reinforce perceptual-motor abilities as well as giving students practice solving basic math problems. The best time to introduce *Designs in Math* is after children completely understand the book's particular math function and need practice solving problems.

It is important that you do not take for granted that students know how to properly use a ruler or straight edge without frustration. Many children have a great deal of difficulty with this skill because they are not properly instructed or do not have ample opportunity to practice. If this is the case, it might be to your students' benefit to begin with the less complicated books in the *Creating Line Designs* series. You may also want to read the "Tips On Using a Straight Edge" page in this book to help in the instruction of drawing with a straight edge.

See the back page of this book for more information on the *Creating Line Designs* series. If the back page is missing, contact your local retail outlet for more information. If your retail outlet does not carry the series you may write to G.E.C. at the address below, or call 1-800-800-1791 for your free catalog and ordering information.

● ●

This Book Develops

- **Visual Discrimination**
- **Right Brain Functioning**
- **Following Directions Skills**
- **Fine Muscle Control**
- **Basic Design Principles**
- **Eye–Hand Coordination**
- **Basic Math Fact Computation/Memorization**

ISBN 1–56500–007–2

Tips on Using a Straight Edge

In teaching children how to use a ruler or straight edge, it is important that the following points be included:

1. The child should hold his hand out with the palm down on the table.

2. He should hold the straight edge with the tips of his thumb and fingers — as far apart as is comfortable. (Note: Most children will hold it in the middle, and the ends will move as they draw their lines.)

3. The child should hold his pencil at a perpendicular angle. (Note: Many children will place the pencil point at a 45–30° angle into the edge of the ruler and usually lift the ruler up with the point of the pencil.)

4. Always have the child move his pencil from top to bottom and from the left to the right. (Left-handed children would move from top to bottom and then right to left.) Instruct him to pick his pencil up and place it on the top or left-side dot after the ruler is in place.

5. After the child has his pencil on one of the dots, he can slide the straight edge to the dot he wants to connect. This is the most accurate way to line up the straight edge with the two dots needing to be connected, and accuracy is an important element in the child's feeling successful about the design he creates.

6. Instruct the child to press gently on the pencil. (Note: Most children will want to press down hard, which causes the straight edge to move, as well as hand fatigue.)

7. Some children may reach over the straight edge to connect the dots, using the awkward edge. Instruct them to draw on the right and top edge of the ruler if they are right handed. (A left-handed child should only draw on the left and top edge of the ruler.)

8. To avoid reaching over the ruler, the student may turn his paper, instead of the straight edge, to a comfortable orientation to facilitate his drawing the designs.

9. Some children (usually the younger ones) may find it easier to stand up while working on their designs.

10. The child must always complete connecting the dots in numerical order then alphabetical order. If they do not, the designs will not work and they will get frustrated to no end — but that is part of the learning process.

✎ Please Note: We wrote the above information referring to a male student. We like girls, too. However, we thought it easier reading to use one gender rather than he/she.

Some students may want to do one or more of the following activities.

Suggestions for Additional Activities

1. Children enjoy using felt pens for coloring the designs. However, every so often, challenge the students to experience different media such as colored pencils, chalk, "artista" oils and even water color paints.
2. Most children will want to use many colors or just certain favorites. Encourage each student to experiment with color patterns of only two or three colors — a great deal of color wheel reinforcement is possible.
3. After he colors the completed design, have the child re-draw the lines with a black pen (when the media permits).
4. Have the child cut out completed designs and paste them on colored background paper. He could even frame the finished product(s).
5. After he colors a design, have the child cut out the specific shapes and paste them onto a colored piece of paper in the exact design seen on the respective "correction key." Many of the designs are difficult. Be certain that the child is capable of doing the chosen design so he doesn't get frustrated.
6. After he colors a design, have the child cut out the specific shapes and paste them onto a colored piece of paper to create a unique design of his own.
7. **Introduce students to string art:**
   a. Several of the designs in this book lend themselves to the craft of string art.
   b. Direct the student to choose one of his favorite designs. (Use your judgement as to the difficulty factor involved.) Have the child complete the design on paper. He does not need to color it.
   c. Give the child a square piece of three-quarter inch thick soft wood (at least 8" square). Plywood is inexpensive and easy to use. It also looks nice when painted or stained a color.
   d. Tape his completed design to the wood with masking tape.
   e. Choose the appropriate nails to work with. (1"–1½" long work well)
   f. Now, direct the student to hammer nails so that about half of each nail remains above the board. Make certain that the nails are placed directly over the dots on the paper taped to the wood.
   g. Have the student tie and run string from one nail to the other until the design is completed. (The stringing may have to be altered from the exact design in order to work. However, encourage the child to allow for design alterations in order to make the stringing easier.)
   h. After the stringing is completed, carefully remove the paper taped to the wood and you have a fine example of string art.

8. **Introduce students to nail sculpture:**

   a. Several of the designs in this book lend themselves to the craft of nail sculpture.

   b. Direct the student to choose one of his favorite designs. (Use your judgement as to the difficulty factor involved.) Have the child complete the design on paper. He does not need to color it.

   c. Give the child a square piece of soft wood (at least 8" square). It must be at least three-fourths inch thick. However, one and one-half inch thick works very well. Try using a rough-sawn exterior wood such as cedar. (Plywood does not work well for this project unless it is hung indoors.) Staining or painting the wood before it is used is optional.

   d. Tape his completed design to the wood with masking tape.

   e. Choose a variety of different nails and screws to use. Make sure to choose some that will rust and some that will not rust (galvanized do not rust.) Get a variety of lengths and head sizes. Brass screws are expensive, but they add a special dimension to the sculpture when it is completed.

   f. Now, direct the student to hammer the nails (or screws), filling an entire shape within the design. (Screws can be hammered very easily. However, screwing them in is excellent for muscle development. He may need instructions on how to make a starter hole with a nail or "Yankee" screwdriver.) Explain that it is important to put the nails as close together as possible in order to create the design effectively. (Some of the shapes within the design may be omitted or changed as needed.)

   g. It is important for the child to use the same type of nail for each of the specific shapes chosen. However, it is equally important for him to vary the different nails within the whole design. The student should also be instructed to vary the depth he hammers the nails into the wood from shape-to-shape. Graduated depths within the same shape create a very nice result.

   h. After the child has filled the desired shapes (some may be left with just the wood showing), get as much of the paper out as possible and have the child hang his finished sculpture outside to be rusted and changed by the weather. The paper will eventually deteriorate.

9. Some students may want to create their designs and have other students complete them.

Division

Correction Key
Design 1

1. Use this design as a correction key.
2. Allow your students to correct their own work.
   Make a transparency of this design and instruct your students to place the transparency over their completed design for a quick and easy check.

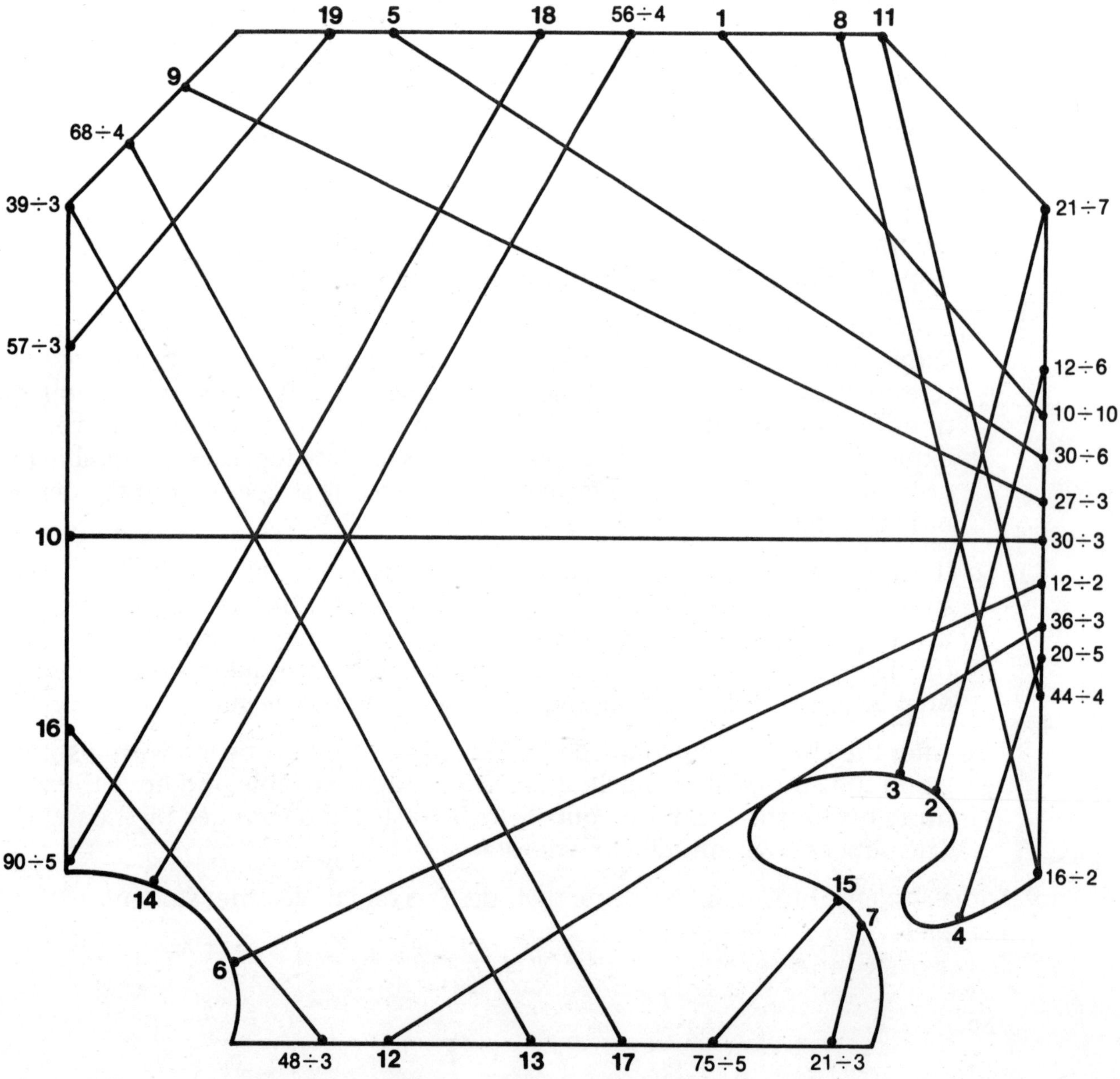

Division
Design 1

Name ______________________

Date ______________________

1. **Read all of the directions before you begin.**
2. **Complete one division problem next to a dot.** (Use scratch paper if you need to.)
3. **After you complete the problem, find the answer on the paper.**
4. **Use a rule (or straight edge) and draw a line to connect the dot by a problem to the dot by the correct answer.**
5. **Complete all of the problems in the same way.**
6. **Carefully color your design after completing all of the problems.**

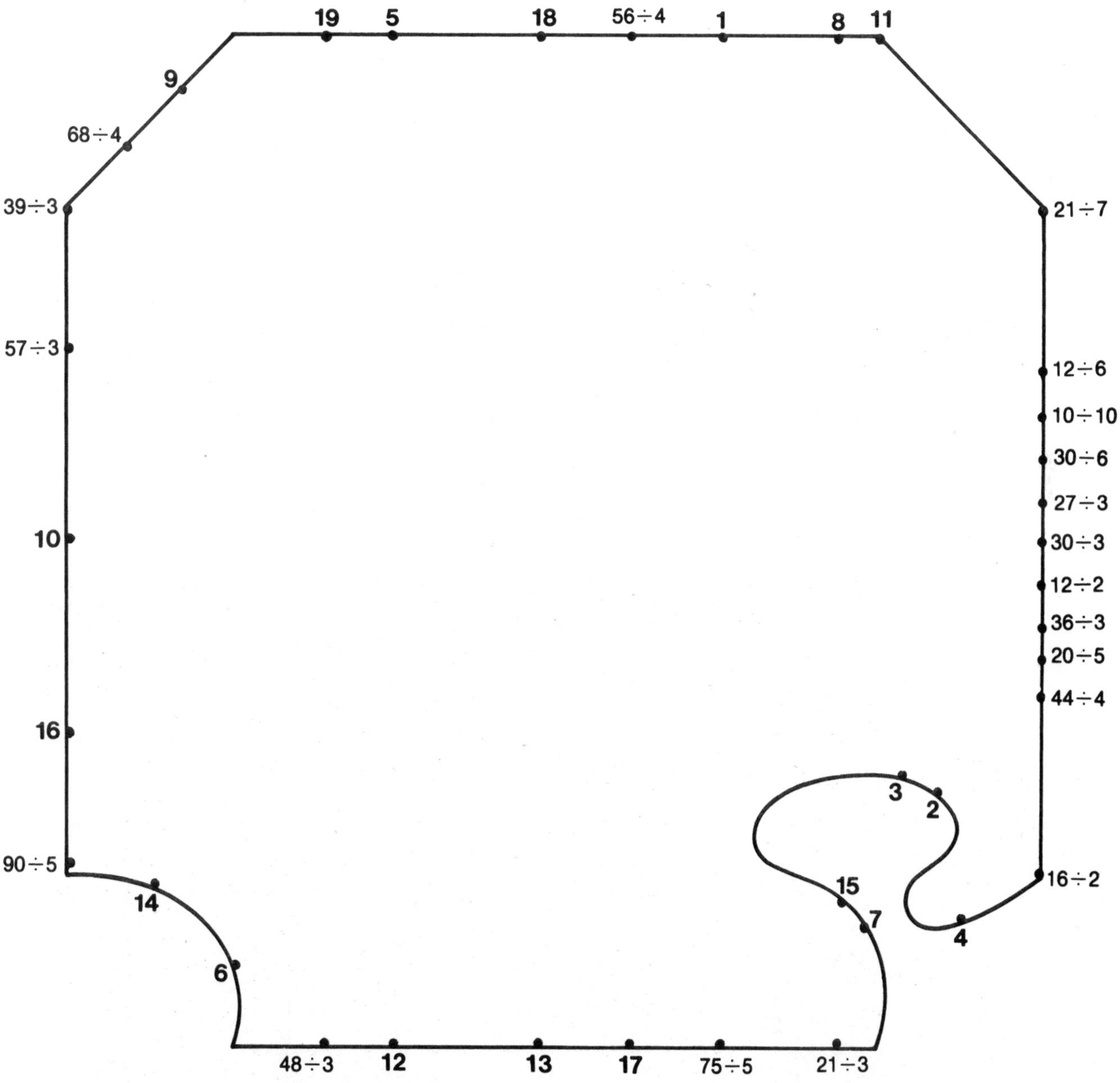

Division

Correction Key
Design 2

1. Use this design as a correction key.
2. Allow your students to correct their own work. Make a transparency of this design and instruct your students to place the transparency over their completed design for a quick and easy check.

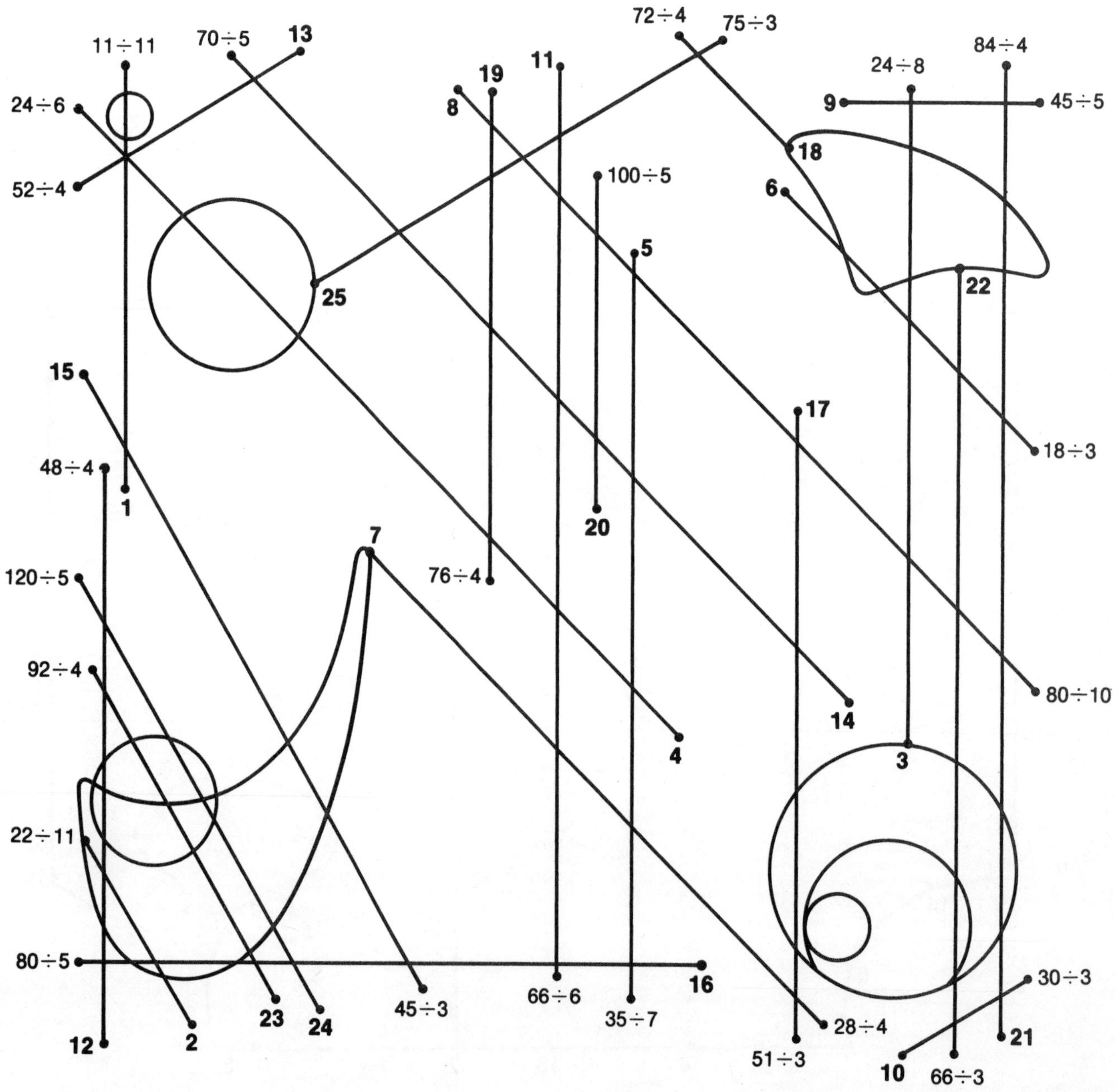

Division

Design 2

Name ______________________

Date ______________________

1. Read all of the directions before you begin.
2. Complete one division problem next to a dot. (Use scratch paper if you need to.)
3. After you complete the problem, find the answer on the paper.
4. Use a rule (or straight edge) and draw a line to connect the dot by a problem to the dot by the correct answer.
5. Complete all of the problems in the same way.
6. Carefully color your design after completing all of the problems.

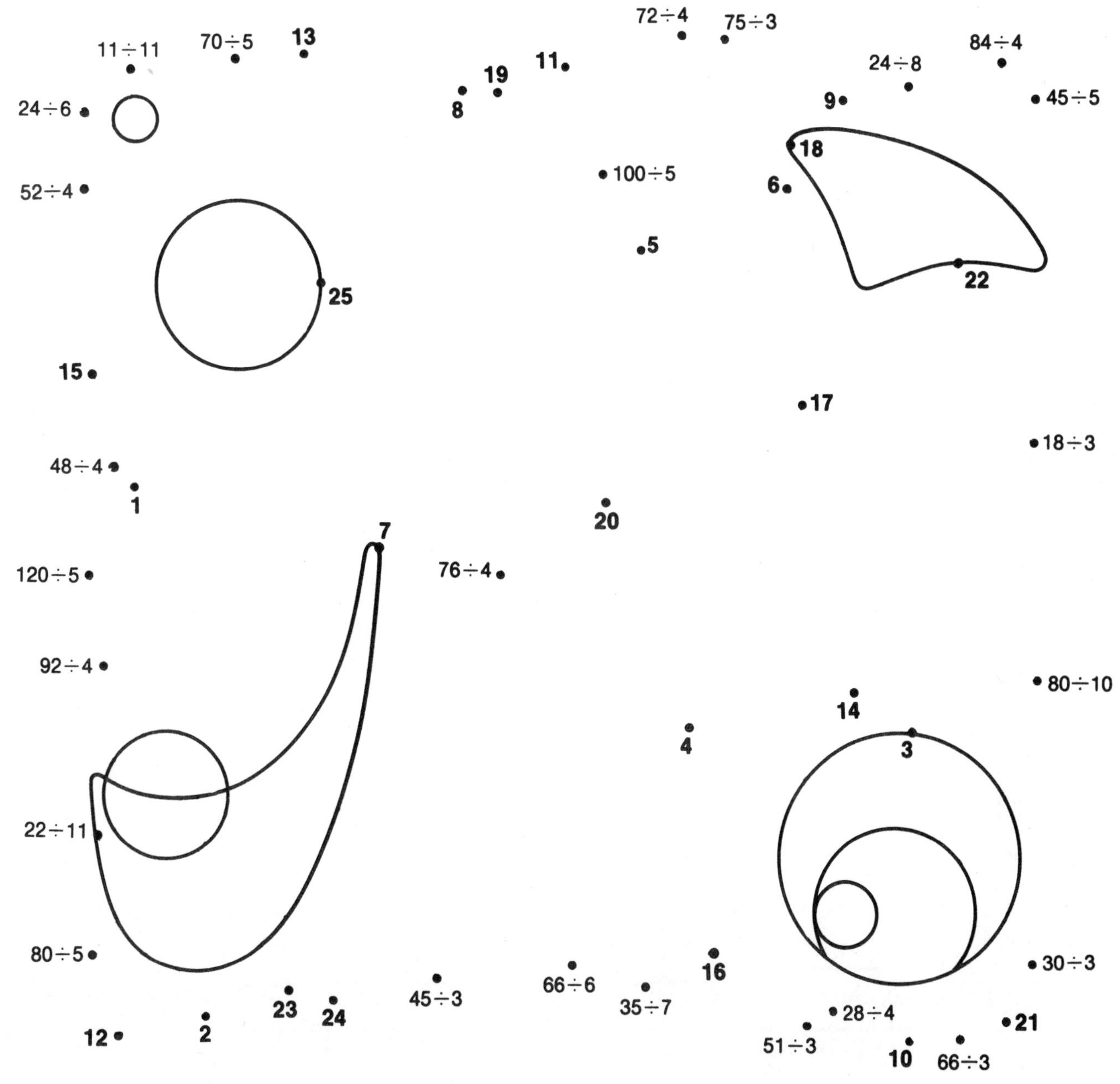

Division

Correction Key
Design 3

1. Use this design as a correction key.
2. Allow your students to correct their own work.
   Make a transparency of this design and instruct your students to place the transparency over their completed design for a quick and easy check.

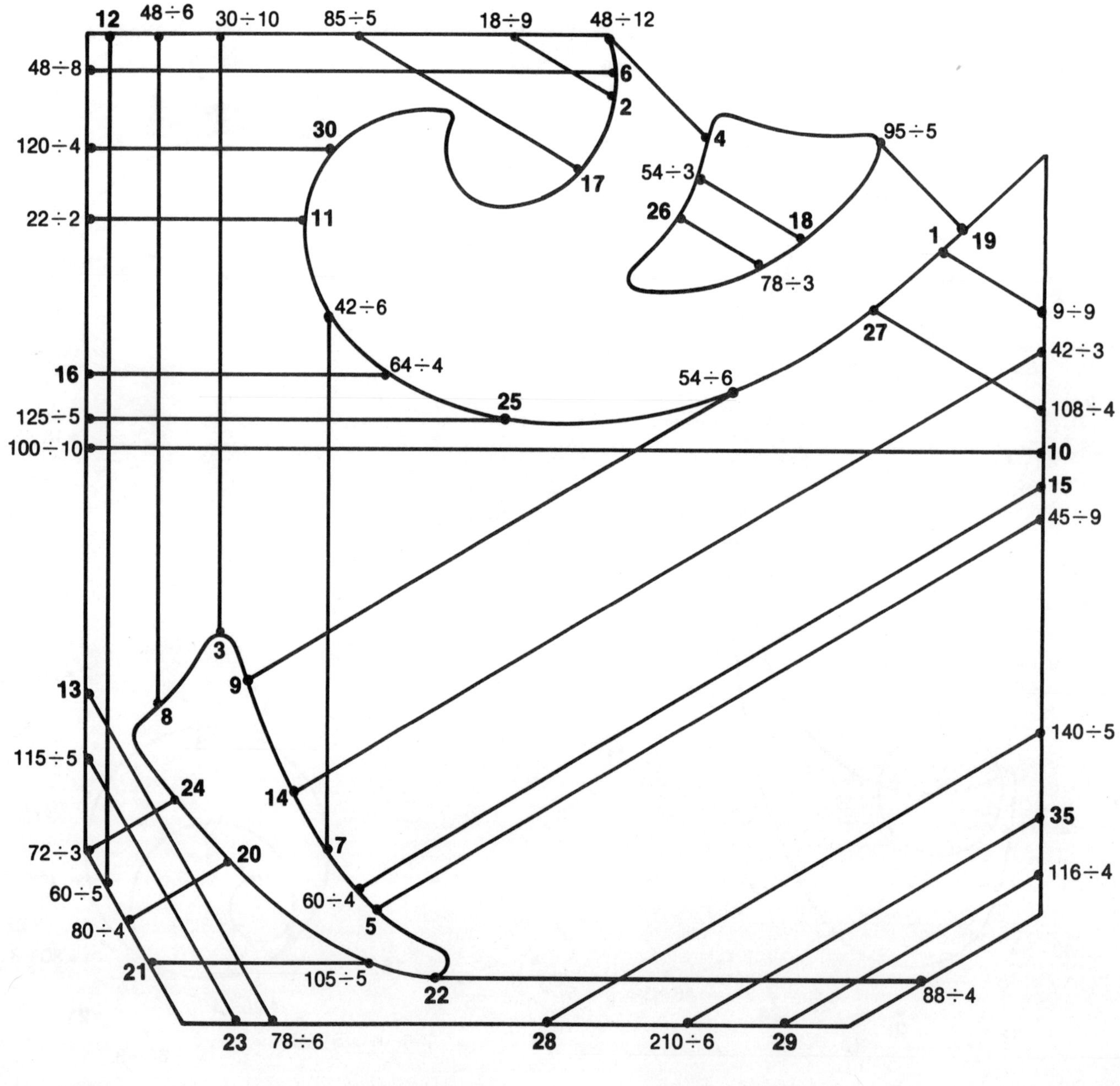

Division
Design 3

Name ______________________

Date ______________________

1. Read all of the directions before you begin.
2. Complete one division problem next to a dot. (Use scratch paper if you need to.)
3. After you complete the problem, find the answer on the paper.
4. Use a rule (or straight edge) and draw a line to connect the dot by a problem to the dot by the correct answer.
5. Complete all of the problems in the same way.
6. Carefully color your design after completing all of the problems.

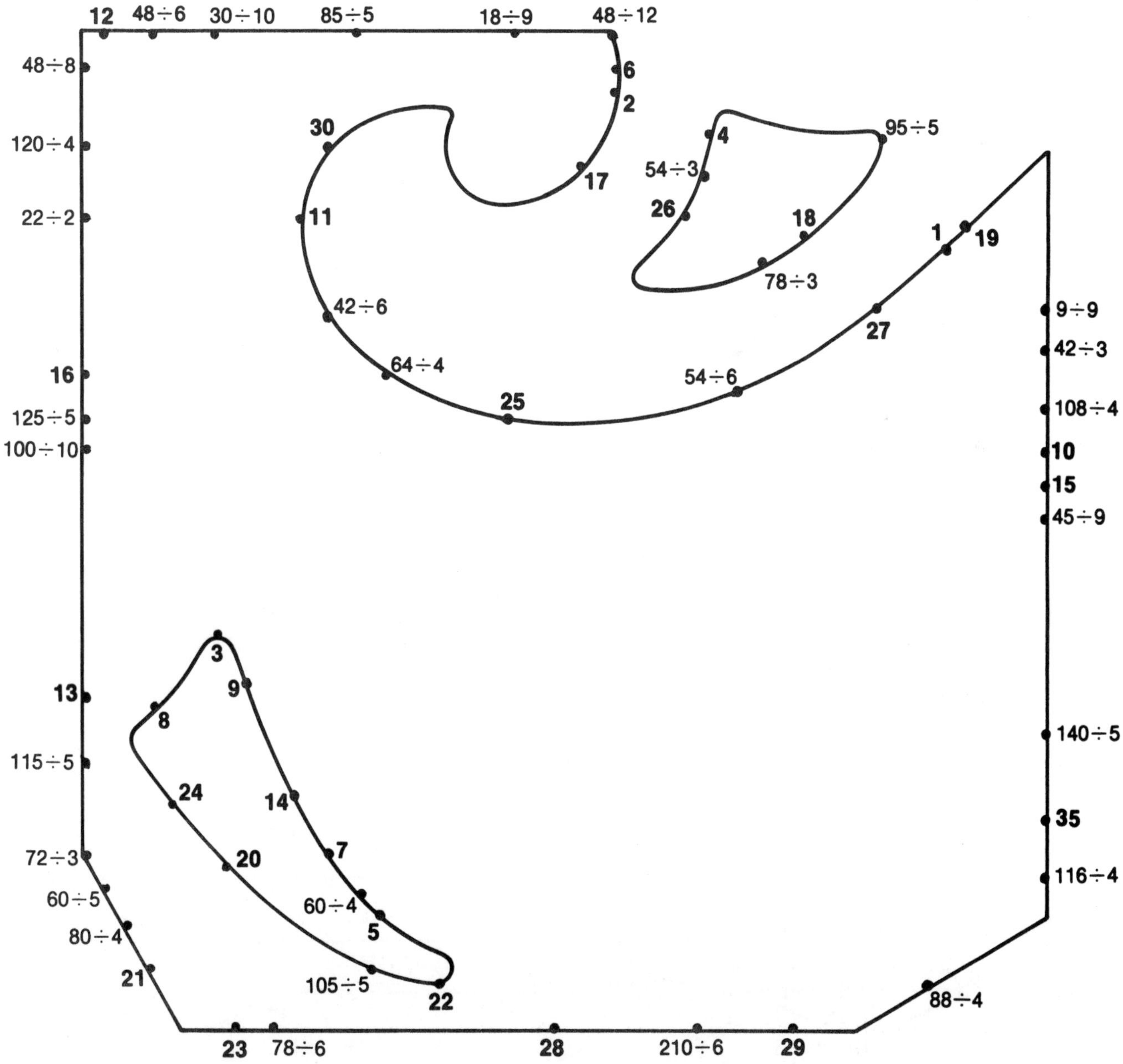

Division

Correction Key
Design 4

1. Use this design as a correction key.
2. Allow your students to correct their own work.
   Make a transparency of this design and instruct your students to place the transparency over their completed design for a quick and easy check.

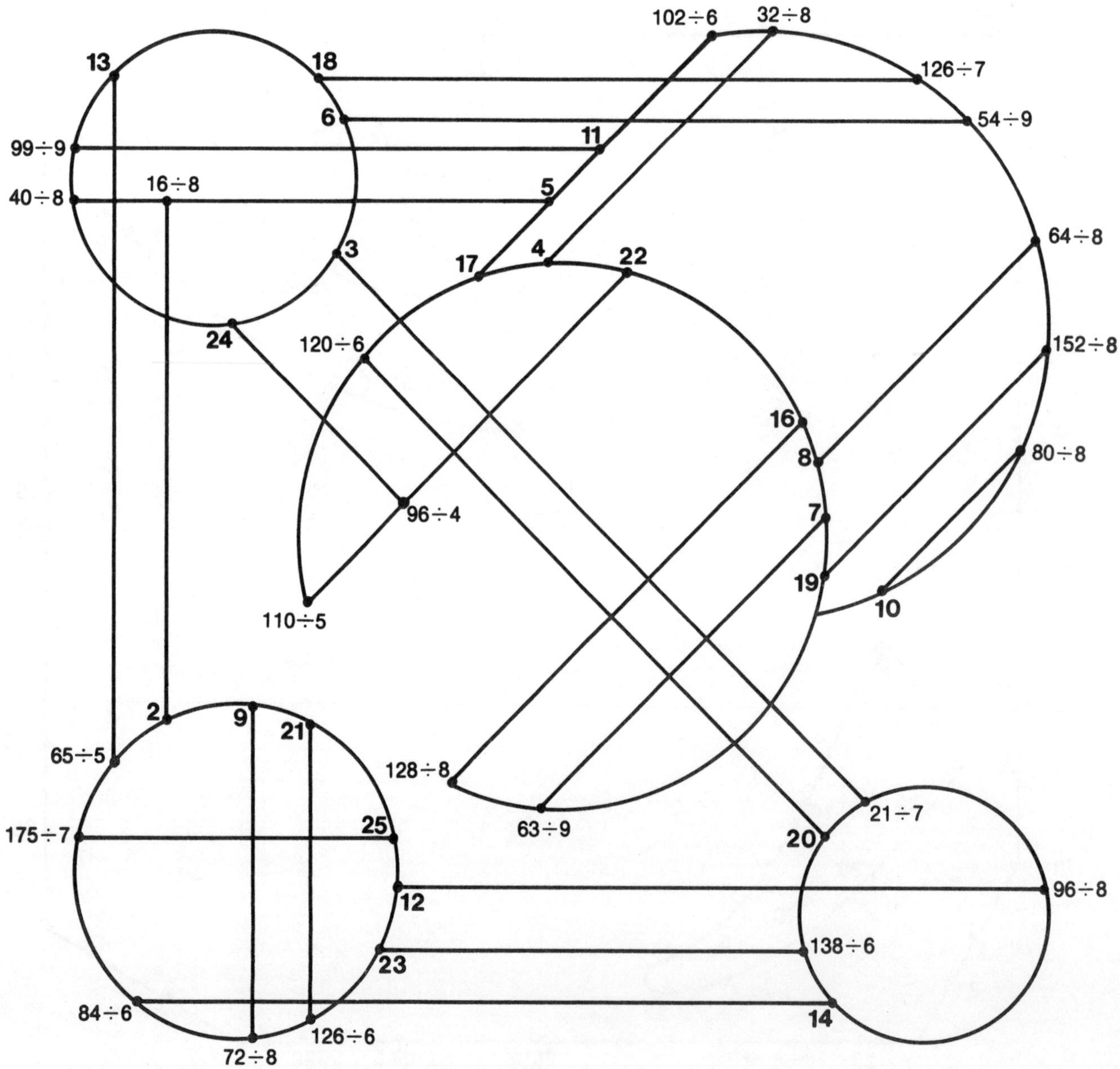

Division
Design 4

Name ______________________

Date ______________________

1. Read all of the directions before you begin.
2. Complete one division problem next to a dot. (Use scratch paper if you need to.)
3. After you complete the problem, find the answer on the paper.
4. Use a rule (or straight edge) and draw a line to connect the dot by the problem to the dot by the correct answer.
5. Complete all of the problems in the same way.
6. Make sure you do all ofthe problems. Some lines pass through dots that are not part of the problem.
7. Carefully color your design after completing all of the problems.

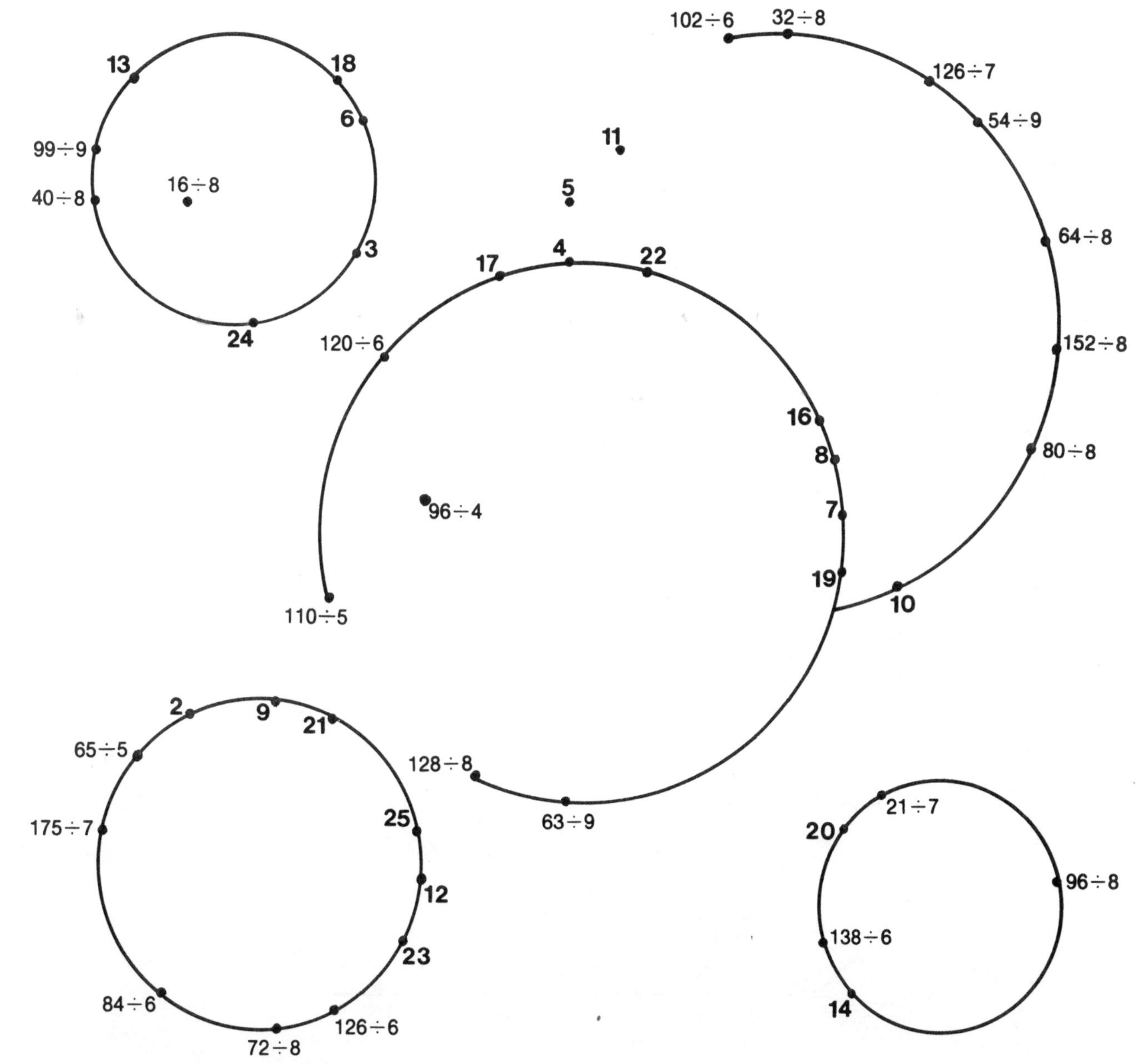

Division

Correction Key
Design 5

1. Use this design as a correction key.
2. Allow your students to correct their own work.
   Make a transparency of this design and instruct your students to place the transparency over their completed design for a quick and easy check.

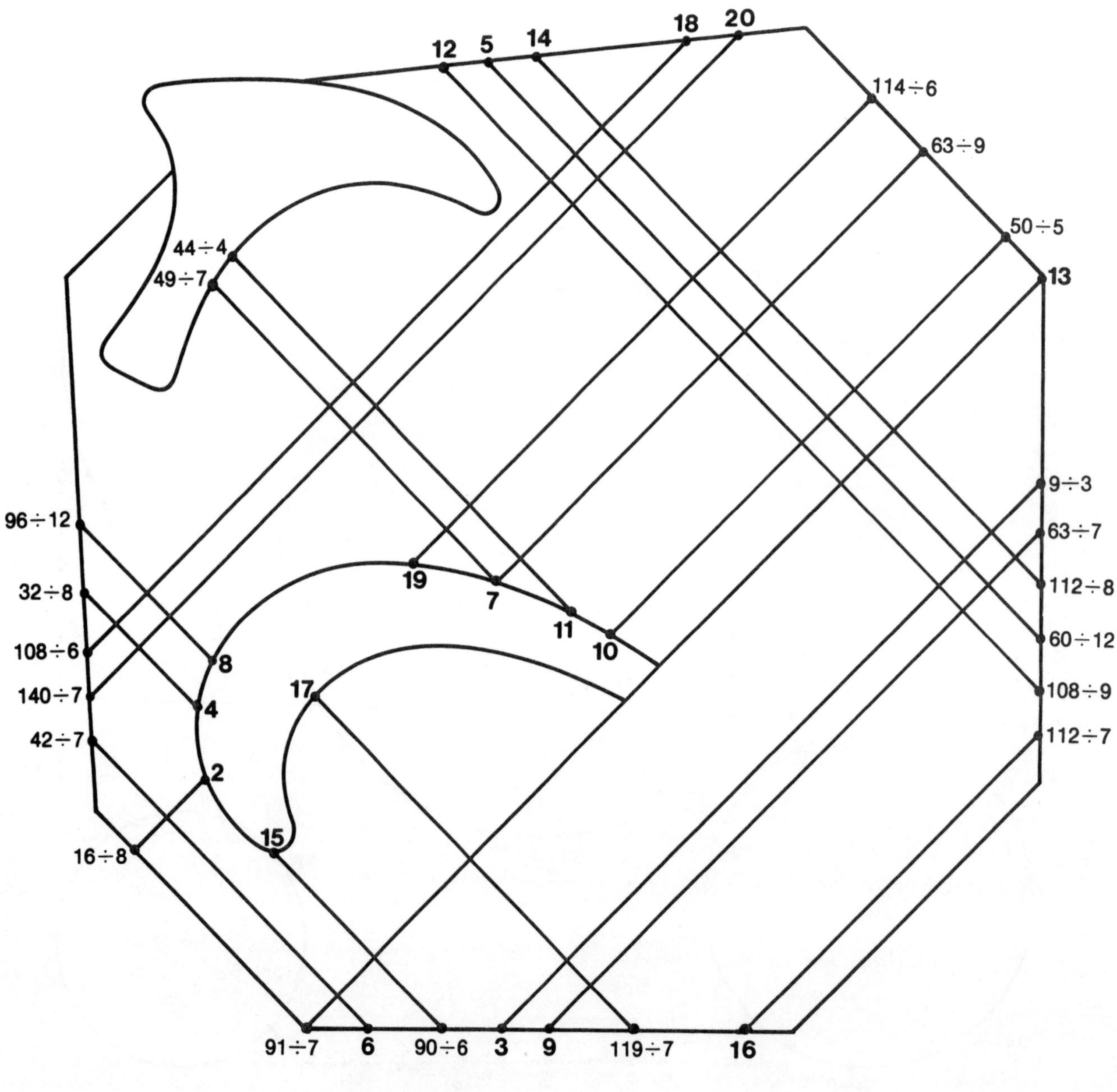

Division
Design 5

Name ______________________

Date ______________________

1. **Read all of the directions before you begin.**
2. **Complete one division problem next to a dot.** (Use scratch paper if you need to.)
3. **After you complete the problem, find the answer on the paper.**
4. **Use a rule (or straight edge) and draw a line to connect the dot by a problem to the dot by the correct answer.**
5. **Complete all of the problems in the same way.**
6. **Make sure you do all of the problems. Some answers are used more than once.**
7. **Carefully color your design after completing all of the problems.**

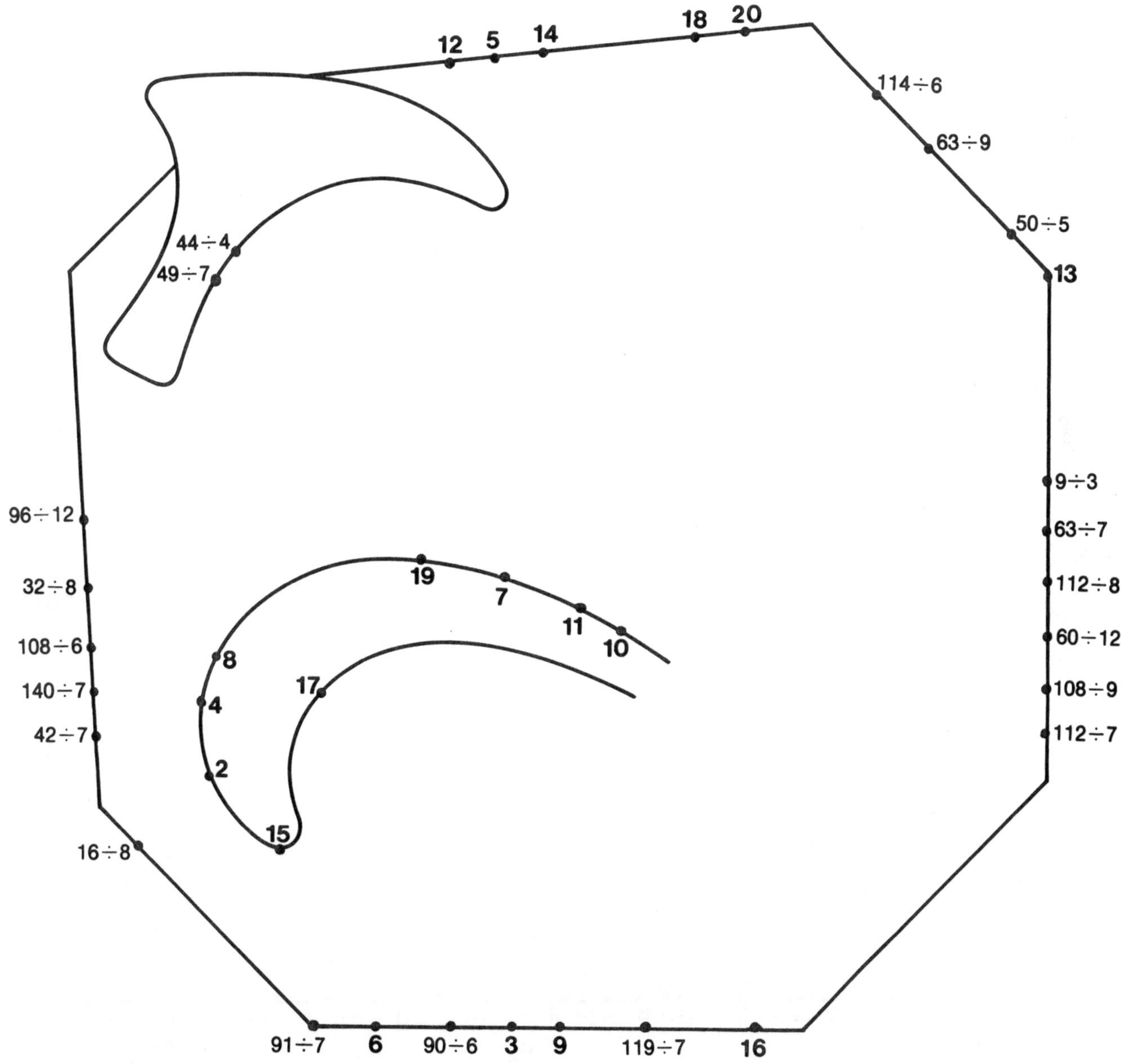

Division

Correction Key
Design 6

1. Use this design as a correction key.
2. Allow your students to correct their own work.
   Make a transparency of this design and instruct your students to place the transparency over their completed design for a quick and easy check.

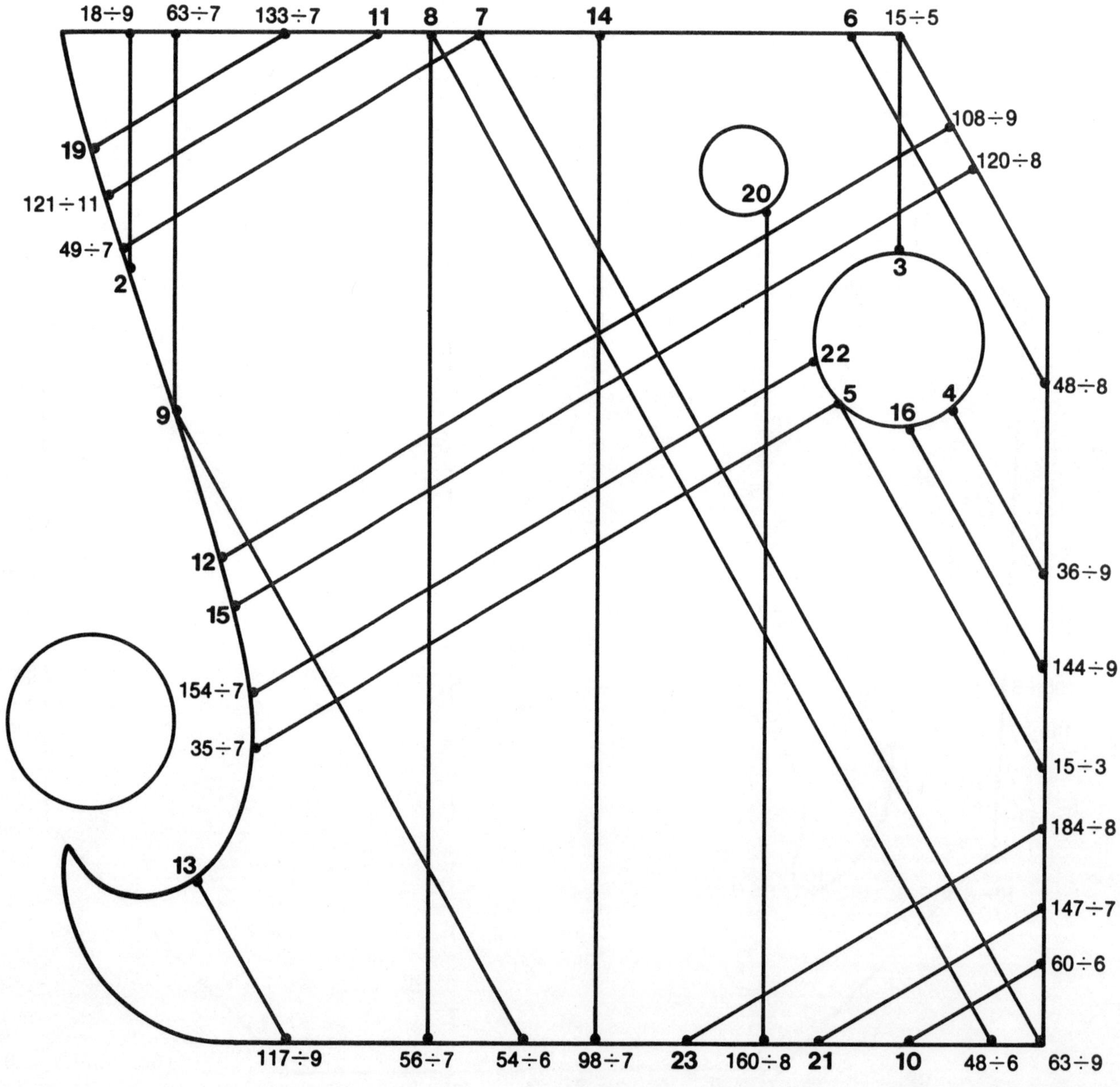

Division
Design 6

Name ______________________

Date ______________________

1. Read all of the directions before you begin.
2. Complete one division problem next to a dot. (Use scratch paper if you need to.)
3. After you complete the problem, find the answer on the paper.
4. Use a rule (or straight edge) and draw a line to connect the dot by a problem to the dot by the correct answer.
5. Complete all of the problems in the same way.
6. Make sure you do all of the problems. Some answers are used more than once.
7. Carefully color your design after completing all of the problems.

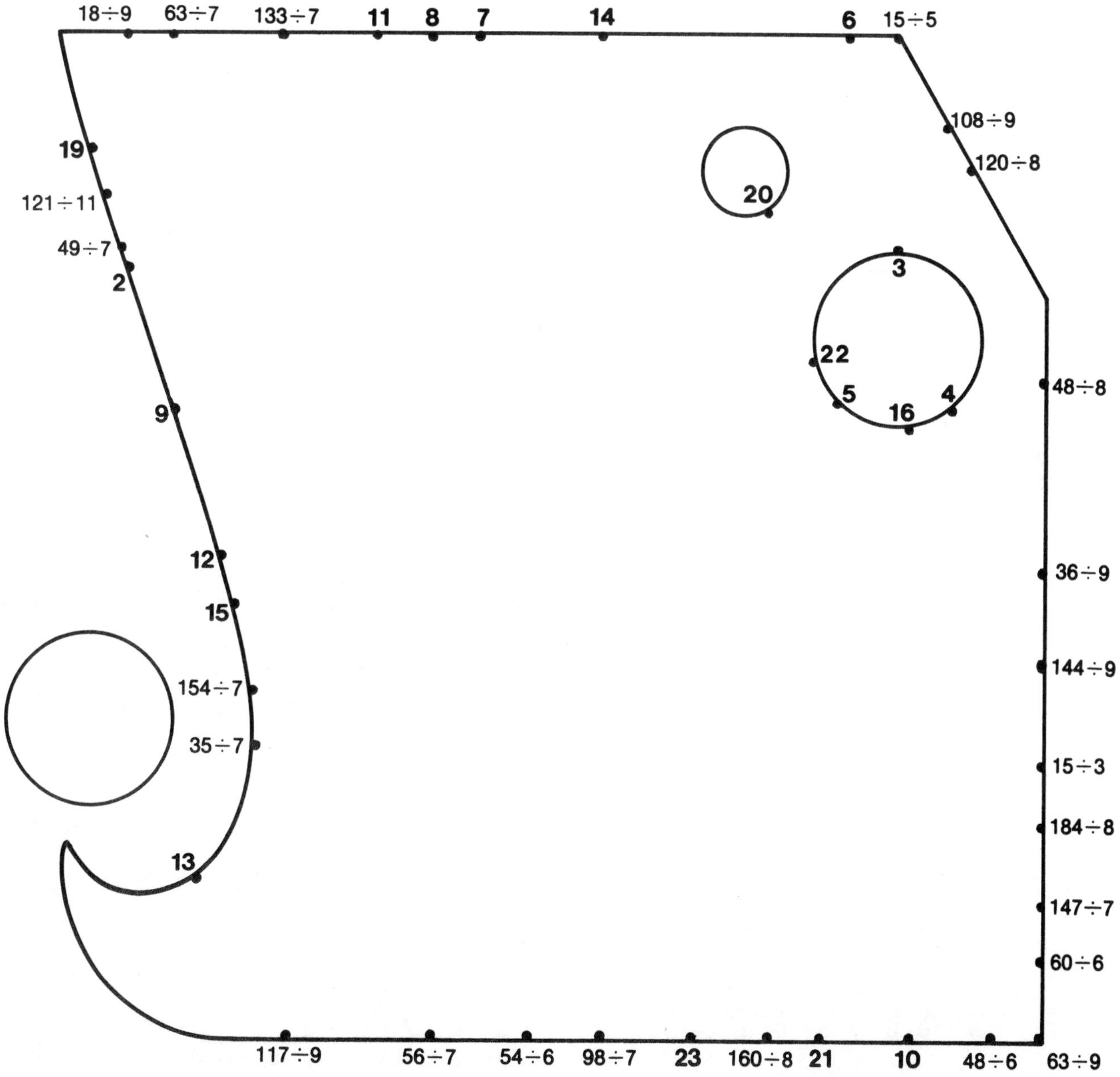

Division

Correction Key
Design 7

1. Use this design as a correction key.
2. Allow your students to correct their own work.
   Make a transparency of this design and instruct your students to place the transparency over their completed design for a quick and easy check.

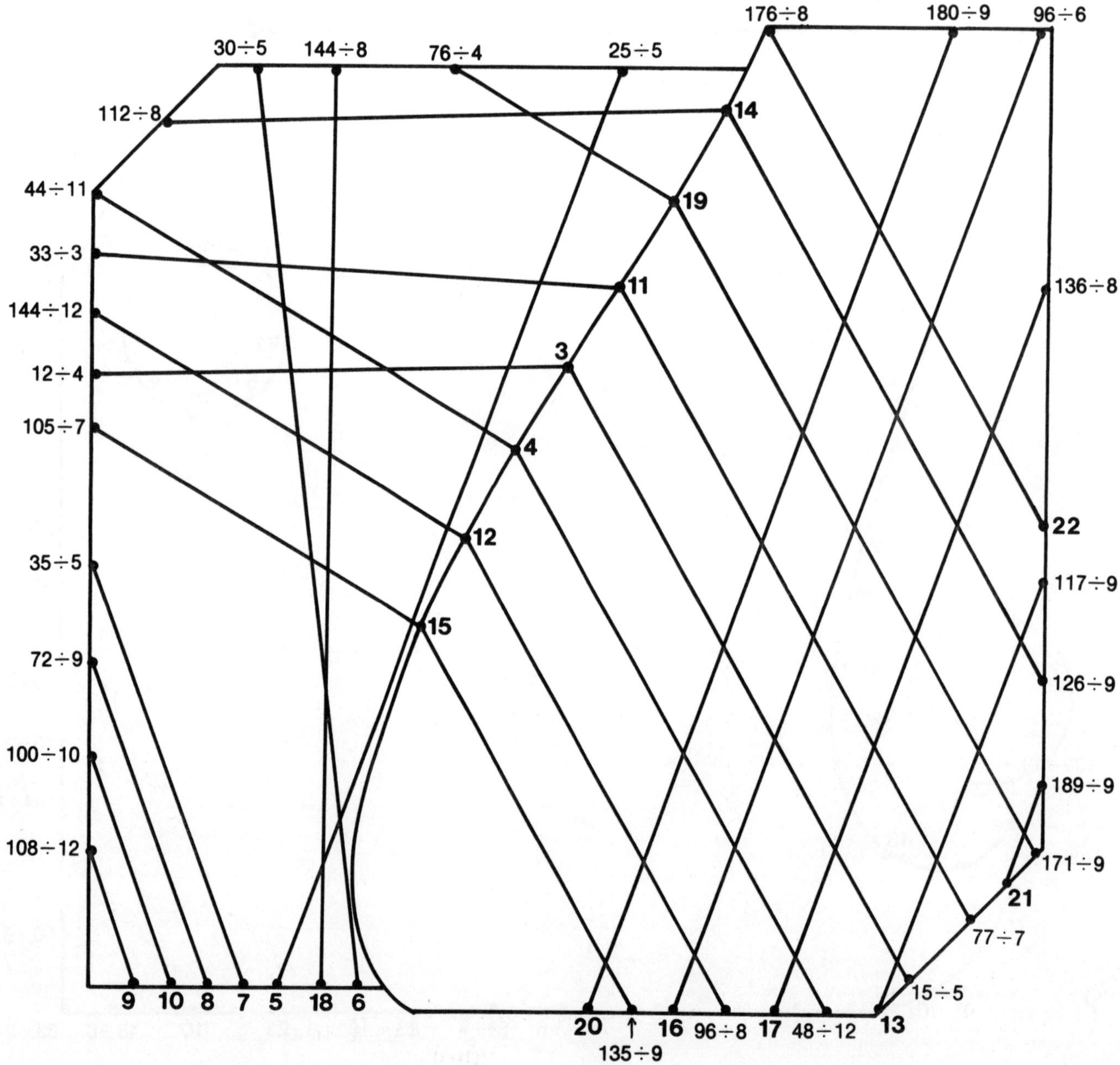

Division
Design 7

Name ____________________

Date ____________________

1. **Read all of the directions before you begin.**
2. **Complete one division problem next to a dot.** (Use scratch paper if you need to.)
3. **After you complete the problem, find the answer on the paper.**
4. **Use a rule (or straight edge) and draw a line to connect the dot by a problem to the dot by the correct answer.**
5. **Complete all of the problems in the same way.**
6. **Make sure you do all of the problems. Some answers are used more than once.**
7. **Carefully color your design after completing all of the problems.**

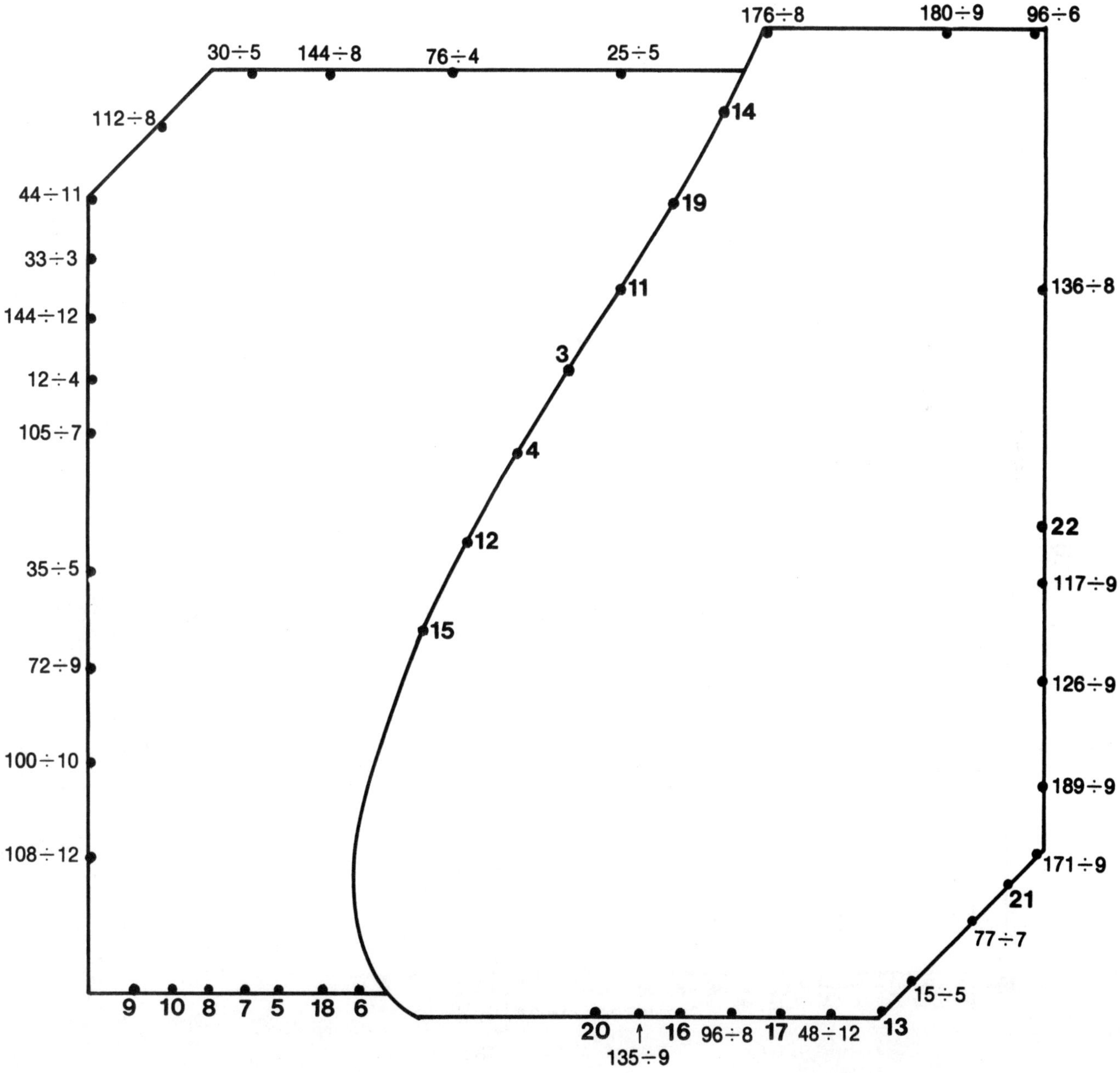

Division

Correction Key
Design 8

1. Use this design as a correction key.
2. Allow your students to correct their own work.
   Make a transparency of this design and instruct your students to place the transparency over their completed design for a quick and easy check.

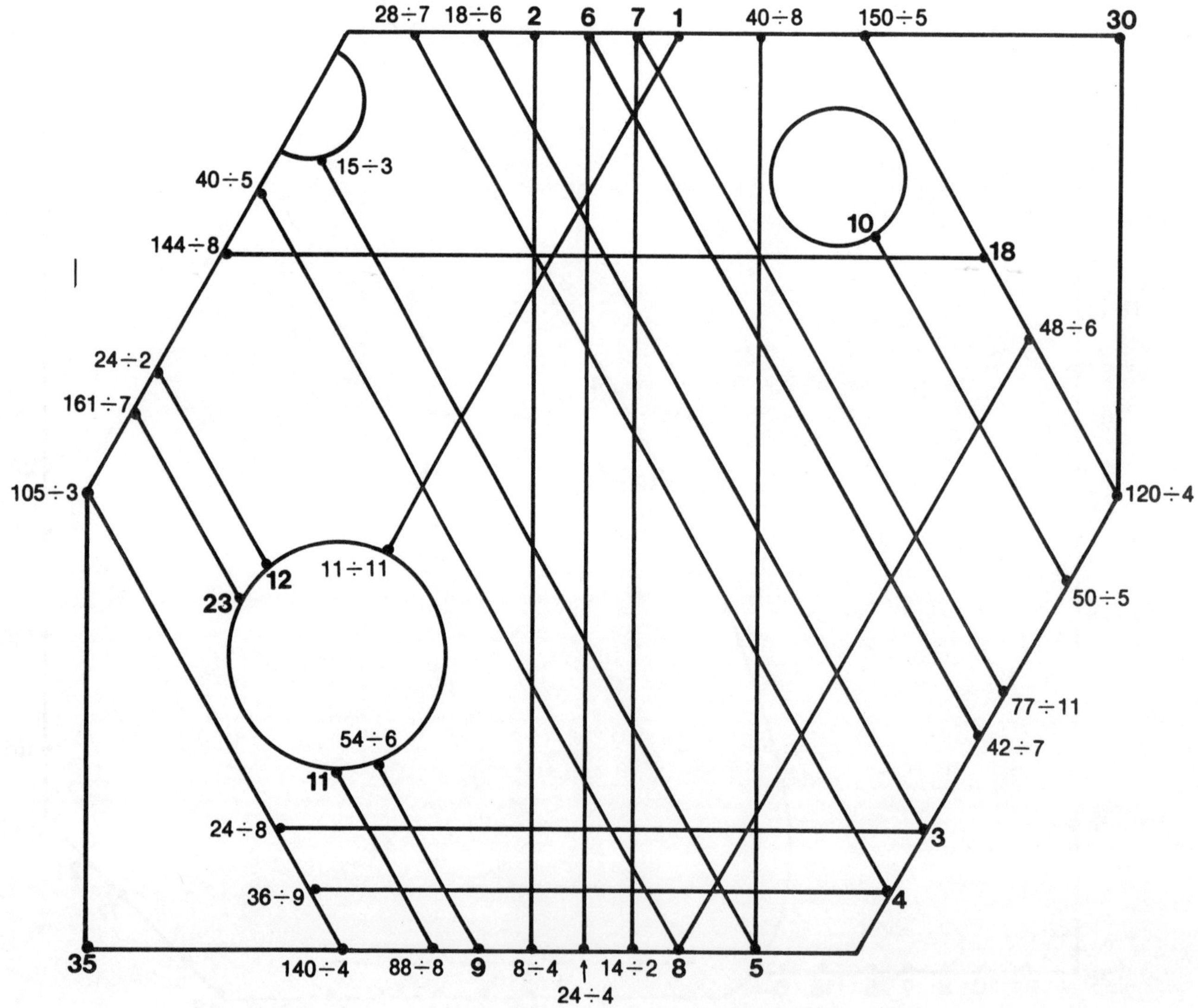

Division
Design 8

Name ______________________

Date ______________________

1. **Read all of the directions before you begin.**
2. **Complete one division problem next to a dot.** (Use scratch paper if you need to.)
3. **After you complete the problem, find the answer on the paper.**
4. **Use a rule (or straight edge) and draw a line to connect the dot by a problem to the dot by the correct answer.**
5. **Complete all of the problems in the same way.**
6. **Make sure you do all of the problems. Some answers are used more than once.**
7. **Carefully color your design after completing all of the problems.**

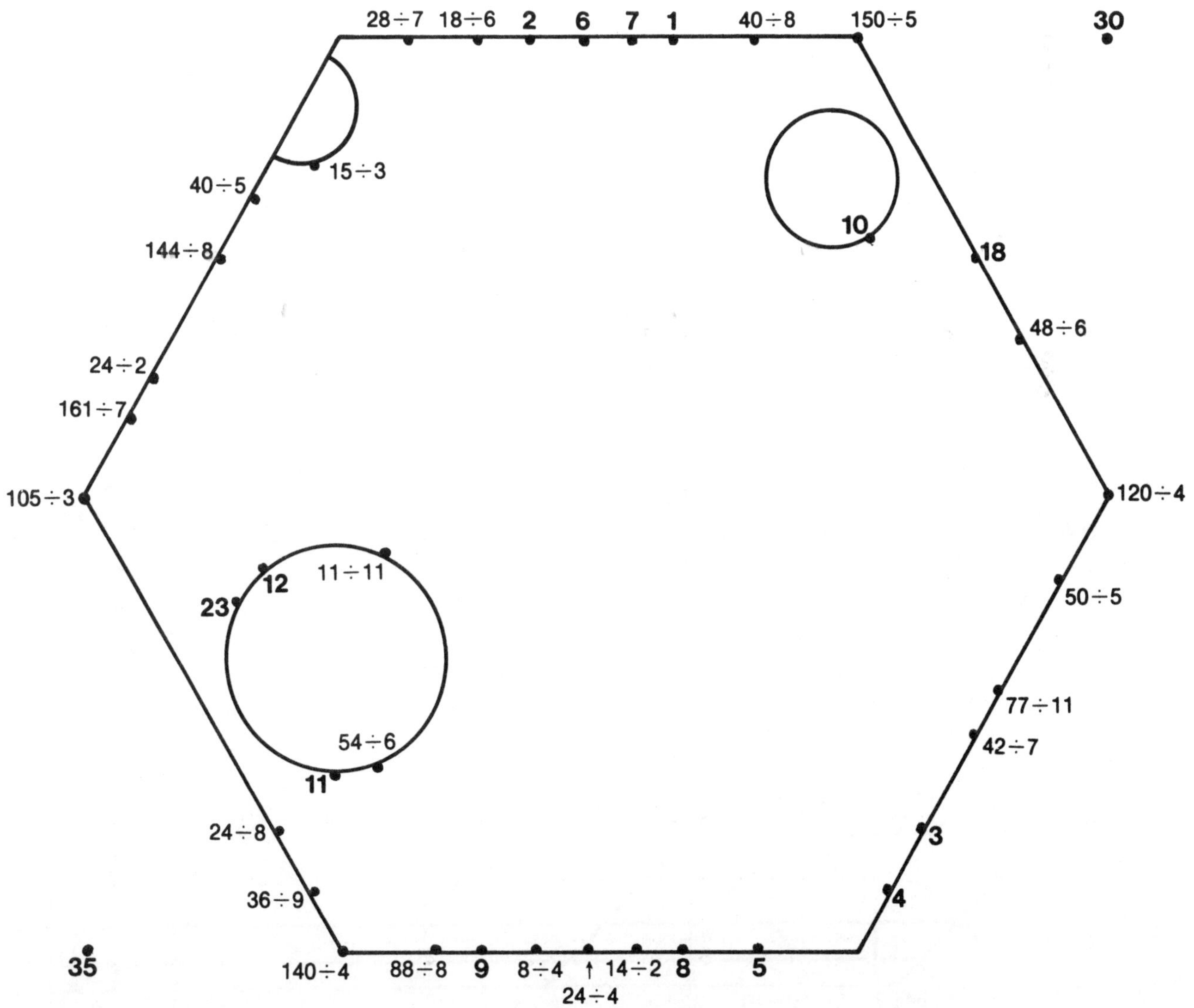

Division

Correction Key
Design 9

1. Use this design as a correction key.
2. Allow your students to correct their own work.
   Make a transparency of this design and instruct your students to place the transparency over their completed design for a quick and easy check.

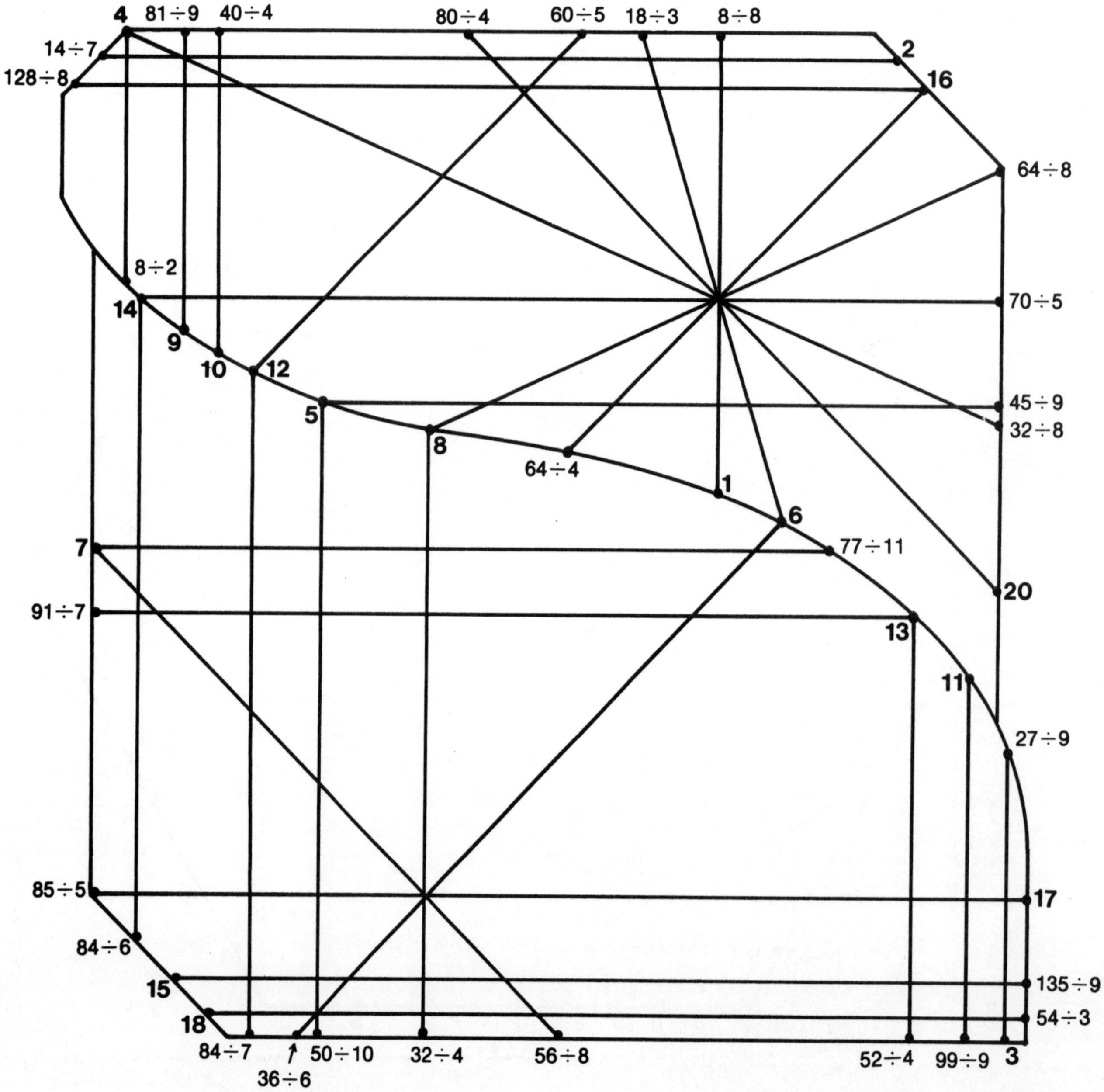

Division

Design 9

Name ______________________

Date ______________________

1. **Read all of the directions before you begin.**
2. **Complete one division problem next to a dot.** (Use scratch paper if you need to.)
3. **After you complete the problem, find the answer on the paper.**
4. **Use a rule (or straight edge) and draw a line to connect the dot by a problem to the dot by the correct answer.**
5. **Complete all of the problems in the same way.**
6. **Make sure you do all of the problems. Some answers are used more than once.**
7. **Carefully color your design after completing all of the problems.**

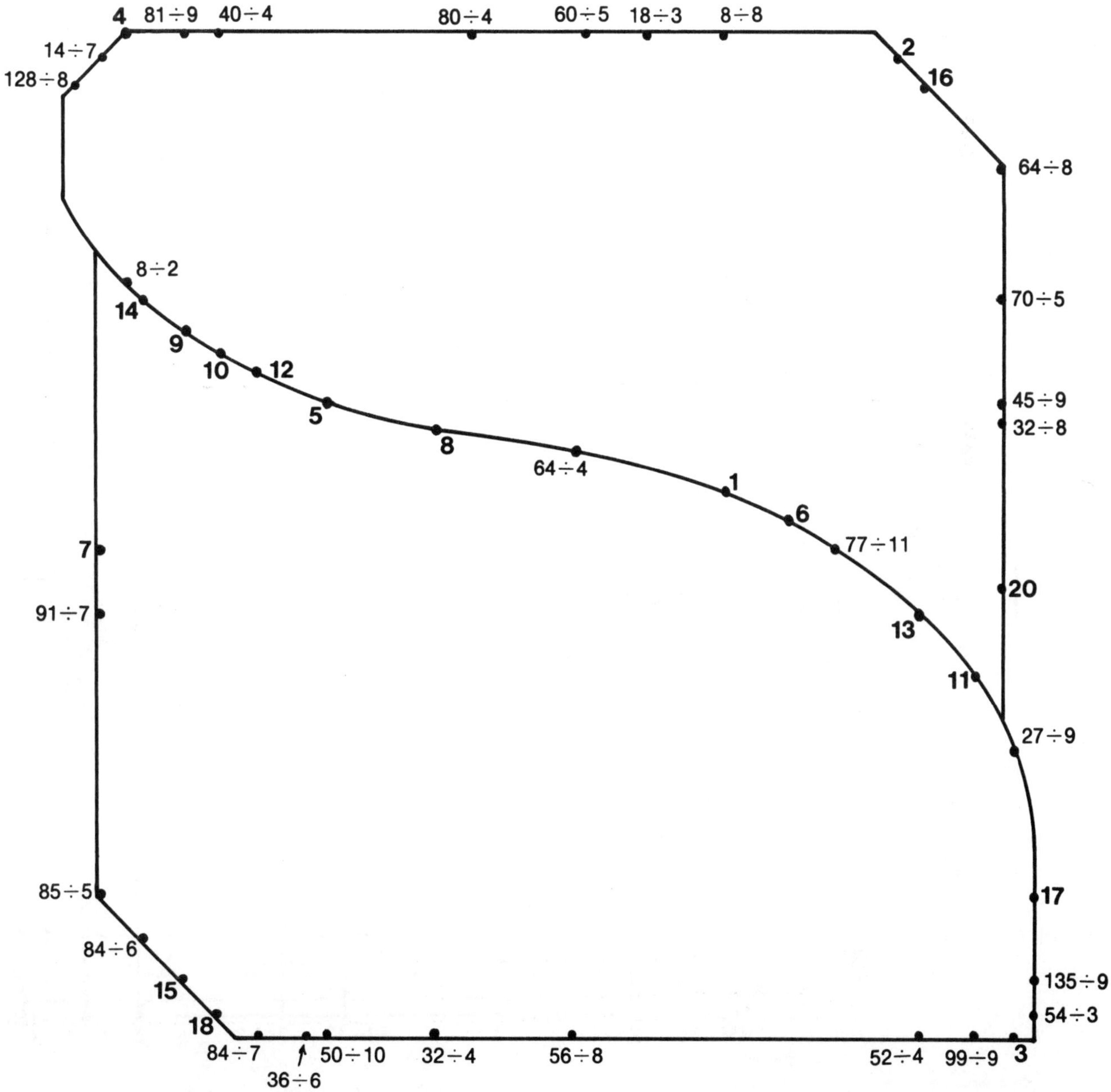

Division

Correction Key
Design 10

1. Use this design as a correction key.
2. Allow your students to correct their own work. Make a transparency of this design and instruct your students to place the transparency over their completed design for a quick and easy check.

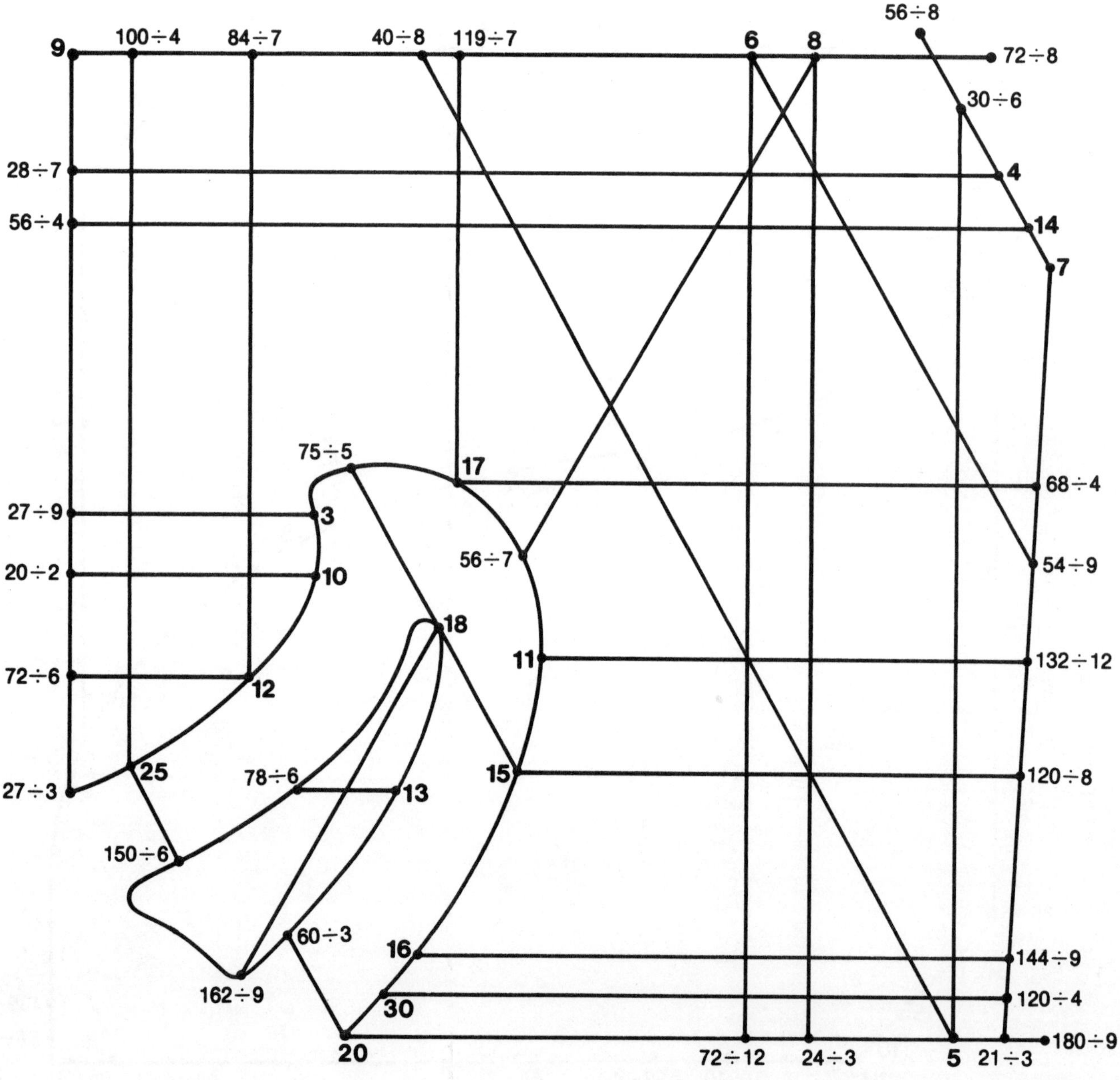

Division
Design 10

Name ______________________________

Date ______________________________

1. Read all of the directions before you begin.
2. Complete one division problem next to a dot. (Use scratch paper if you need to.)
3. After you complete the problem, find the answer on the paper.
4. Use a rule (or straight edge) and draw a line to connect the dot by a problem to the dot by the correct answer.
5. Complete all of the problems in the same way.
6. Make sure you do all of the problems. Some lines pass through dots that are not part of the problem and some answers are used twice.
7. Carefully color your design after completing all of the problems.

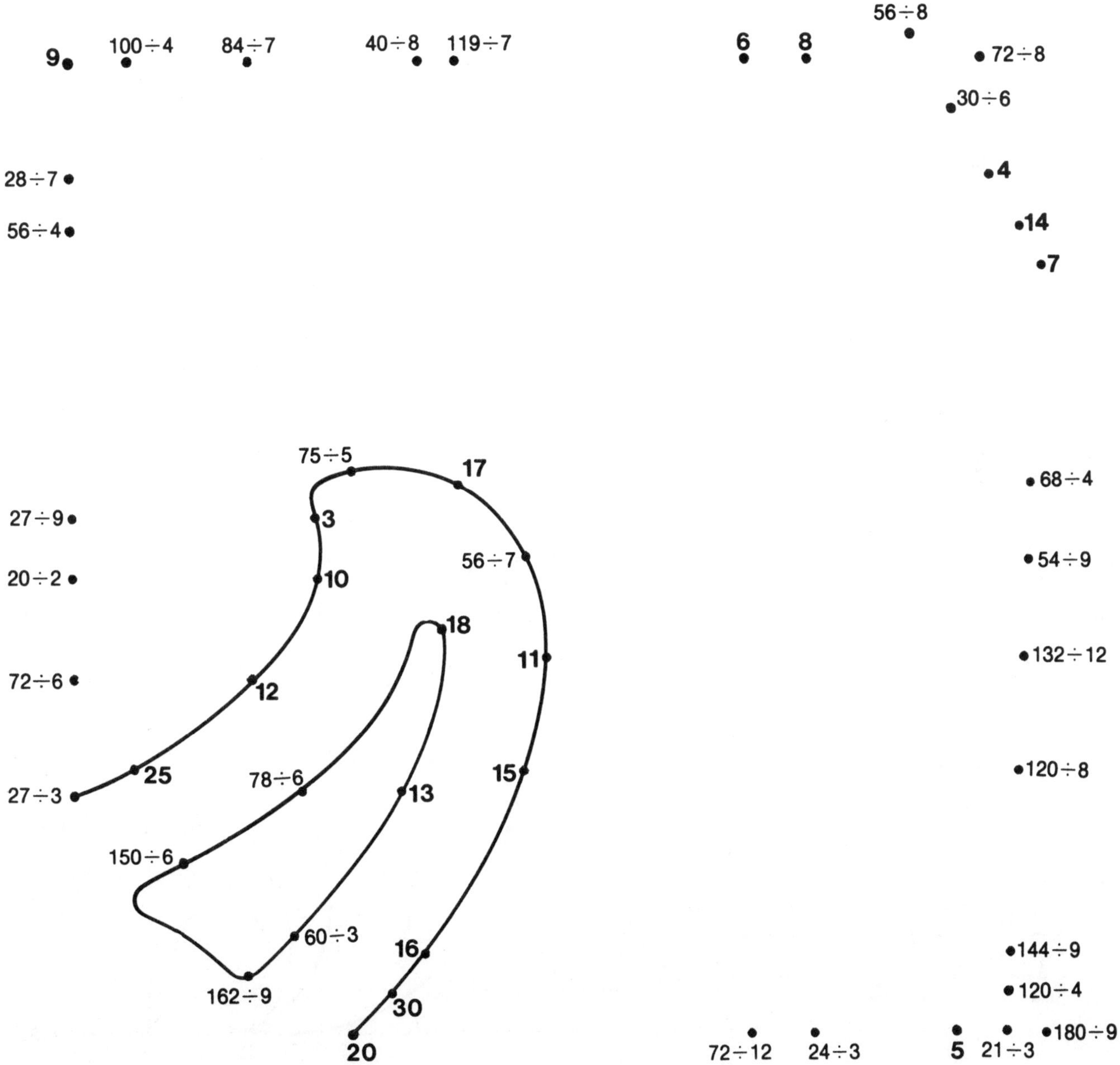

Division

Correction Key
Design 11

1. Use this design as a correction key.
2. Allow your students to correct their own work.
   Make a transparency of this design and instruct your students to place the transparency over their completed design for a quick and easy check.

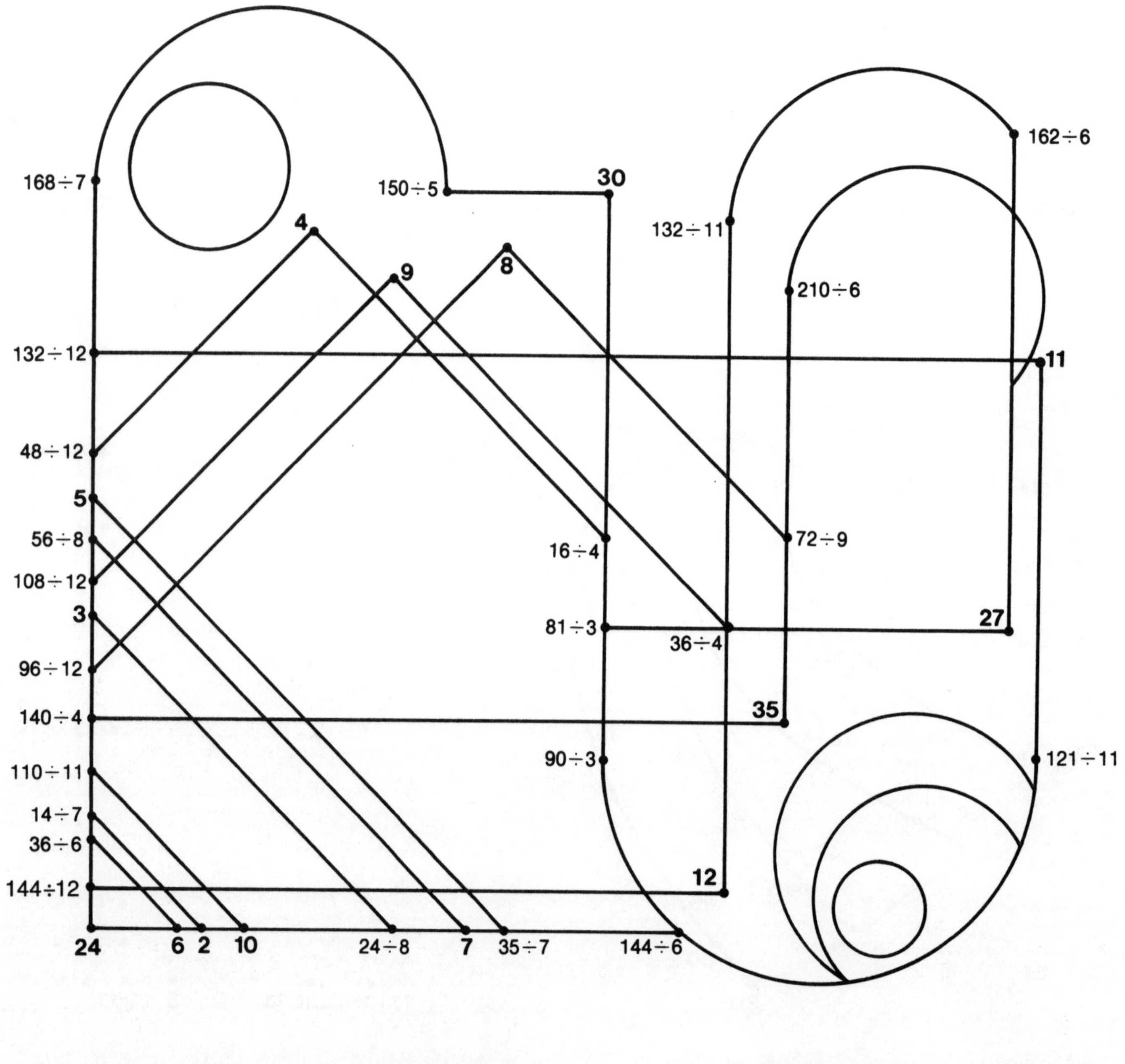

Division
Design 11

Name ______________________

Date ______________________

1. Read all of the directions before you begin.
2. Complete one division problem next to a dot. (Use scratch paper if you need to.)
3. After you complete the problem, find the answer on the paper.
4. Use a rule (or straight edge) and draw a line to connect the dot by a problem to the dot by the correct answer.
5. Complete all of the problems in the same way.
6. Make sure you do all of the problems. Some lines pass through dots that are not part of the problem and some answers are used twice.
7. Carefully color your design after completing all of the problems.

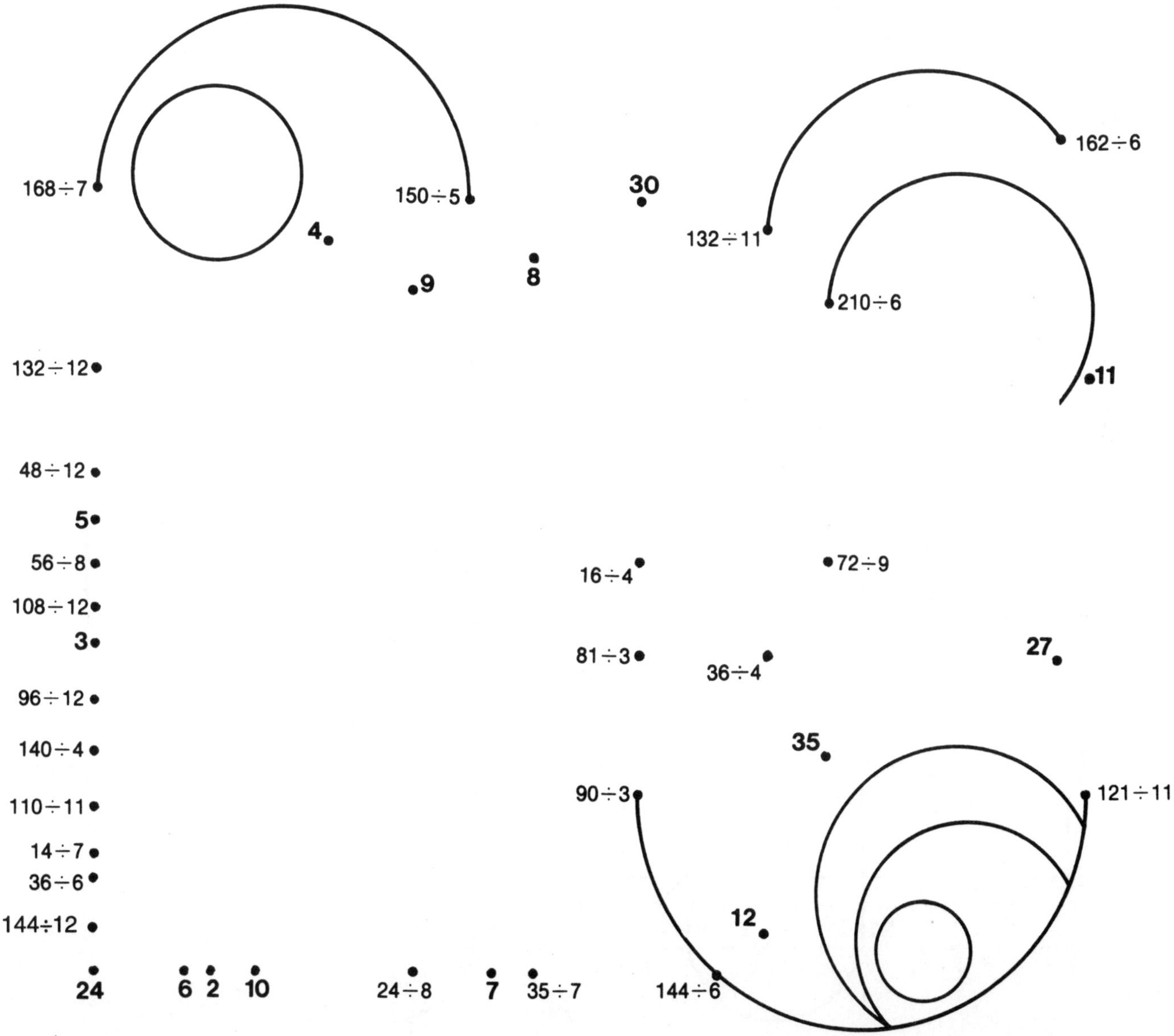

Division

Correction Key
Design 12

1. Use this design as a correction key.
2. Allow your students to correct their own work. Make a transparency of this design and instruct your students to place the transparency over their completed design for a quick and easy check.

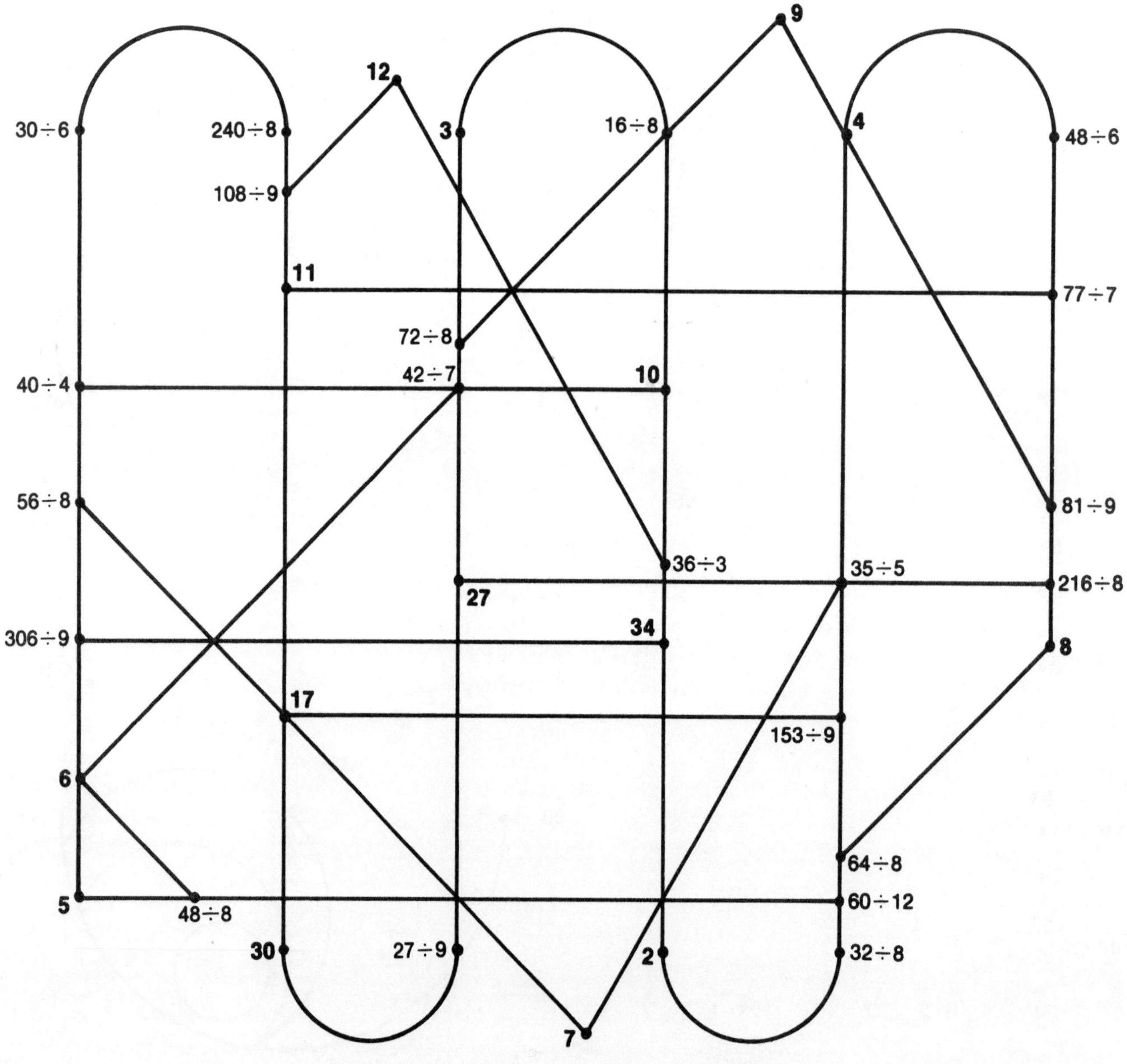

Division
Design 12

Name ______________________

Date ______________________

1. Read all of the directions before you begin.
2. Complete one division problem next to a dot. (Use scratch paper if you need to.)
3. After you complete the problem, find the answer on the paper.
4. Use a rule (or straight edge) and draw a line to connect the dot by a problem to the dot by the correct answer.
5. Complete all of the problems in the same way.
6. Make sure you do all of the problems. Some lines pass through dots that are not part of the problem and some answers are used twice.
7. Carefully color your design after completing all of the problems.

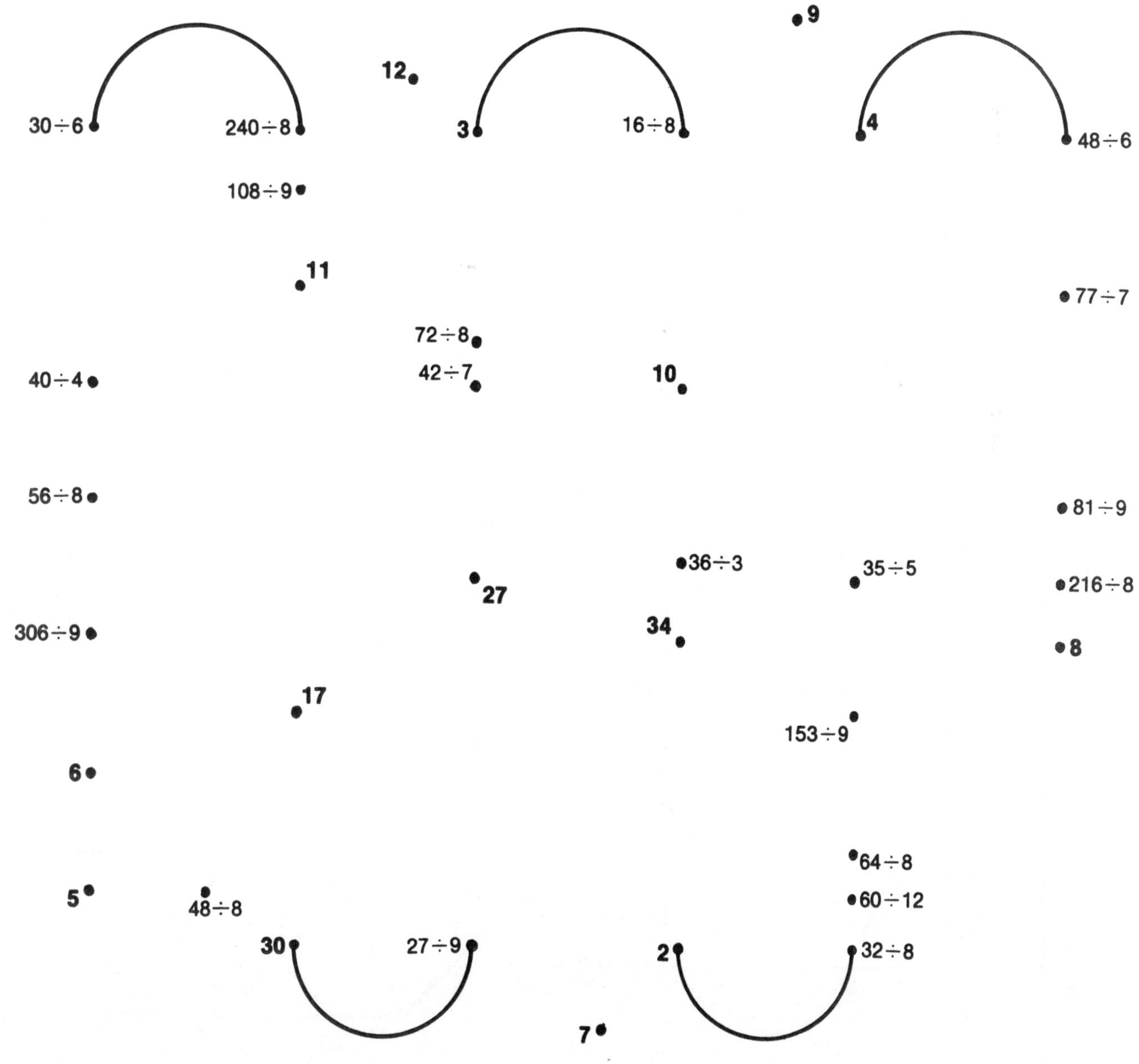

Division

Correction Key
Design 13

1. Use this design as a correction key.
2. Allow your students to correct their own work.
   Make a transparency of this design and instruct your students to place the transparency over their completed design for a quick and easy check.

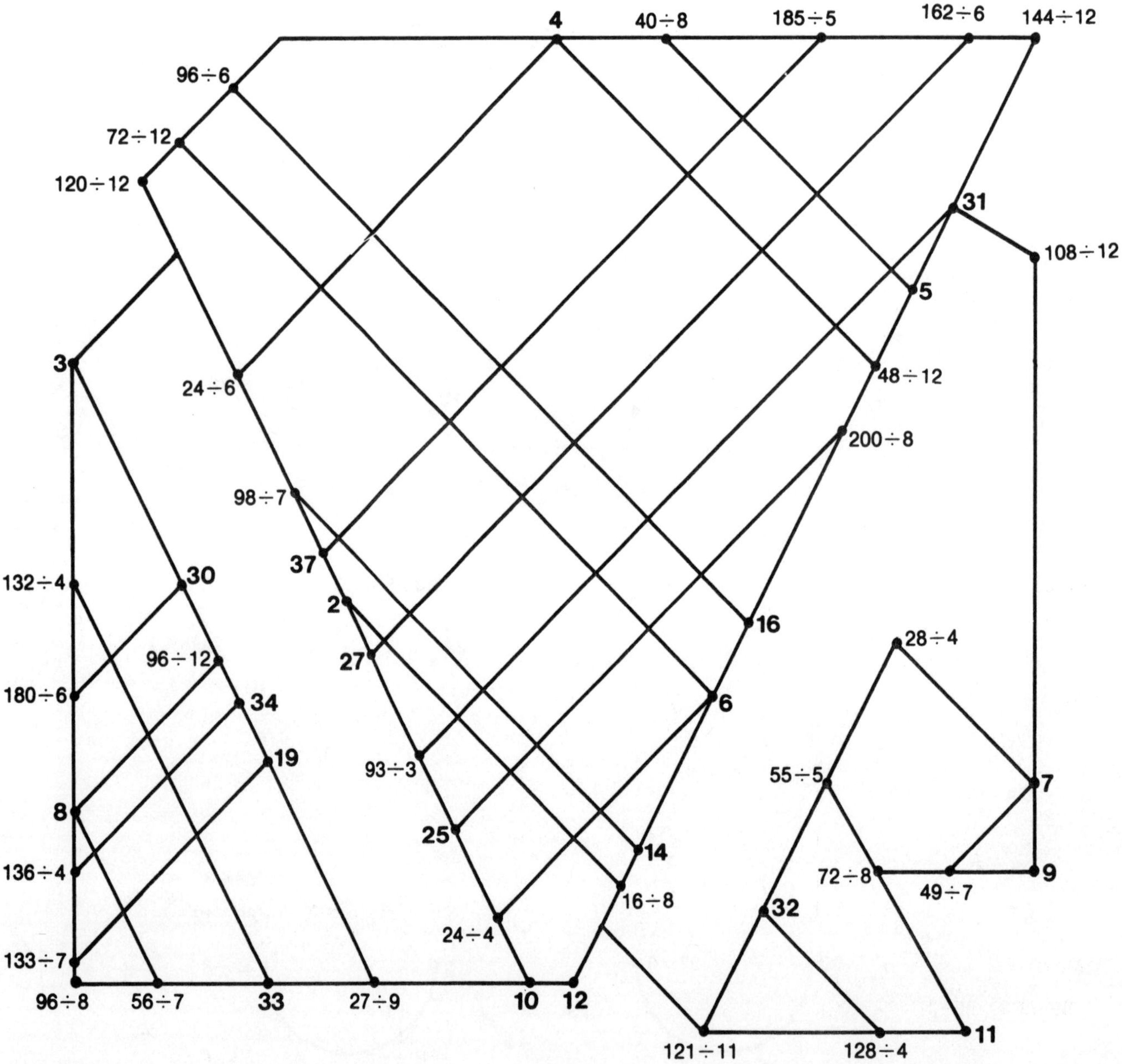

Division
Design 13

Name ______________________

Date ______________________

1. **Read all of the directions before you begin.**
2. **Complete one division problem next to a dot.** (Use scratch paper if you need to.)
3. **After you complete the problem, find the answer on the paper.**
4. **Use a rule (or straight edge) and draw a line to connect the dot by a problem to the dot by the correct answer.**
5. **Complete all of the problems in the same way.**
6. **Make sure you do all of the problems. Some lines pass through dots that are not part of the problem and some answers are used twice.**
7. **Carefully color your design after completing all of the problems.**

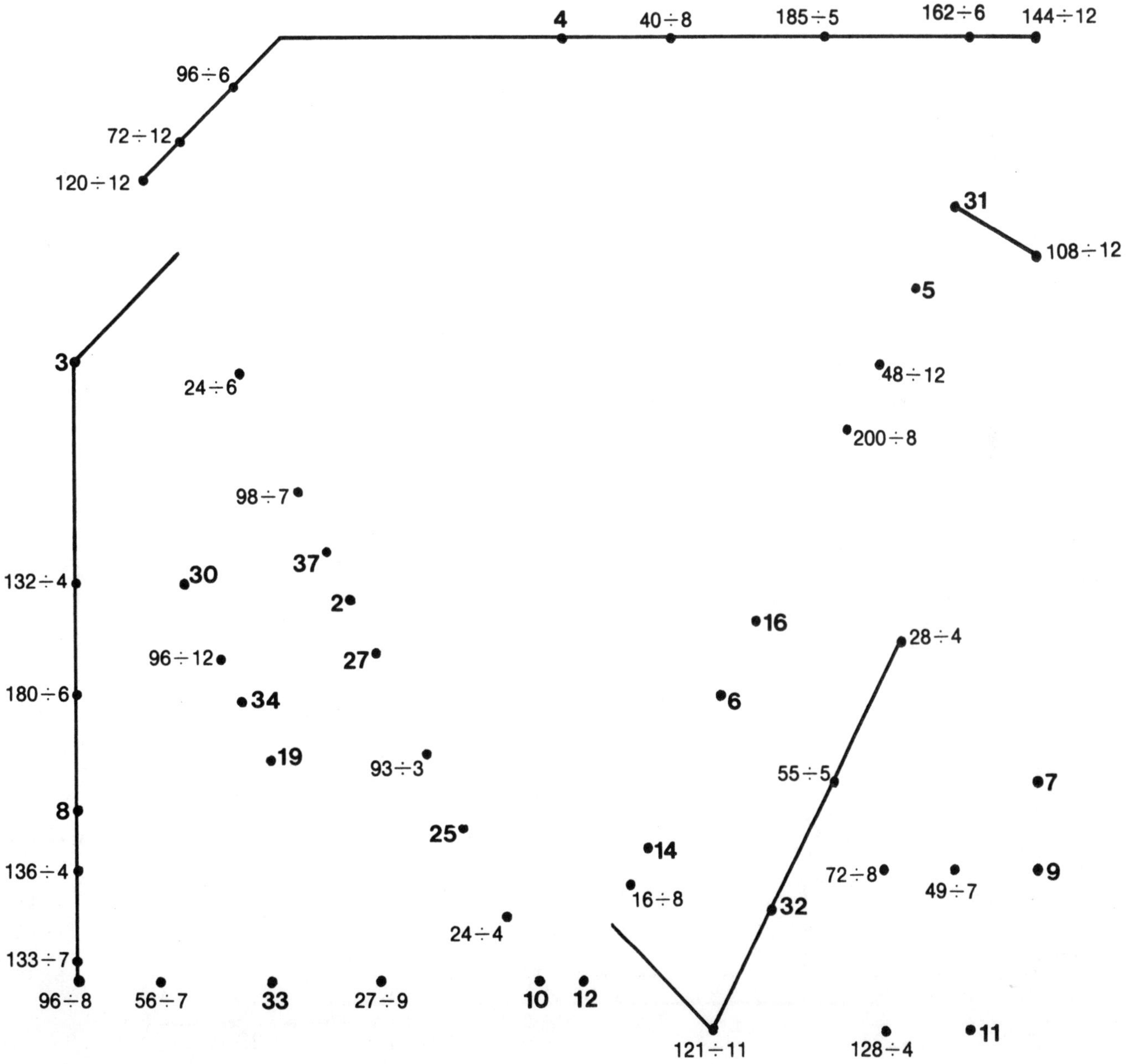

Division

Correction Key
Design 14

1. Use this design as a correction key.
2. Allow your students to correct their own work.
   Make a transparency of this design and instruct your students to place the transparency over their completed design for a quick and easy check.

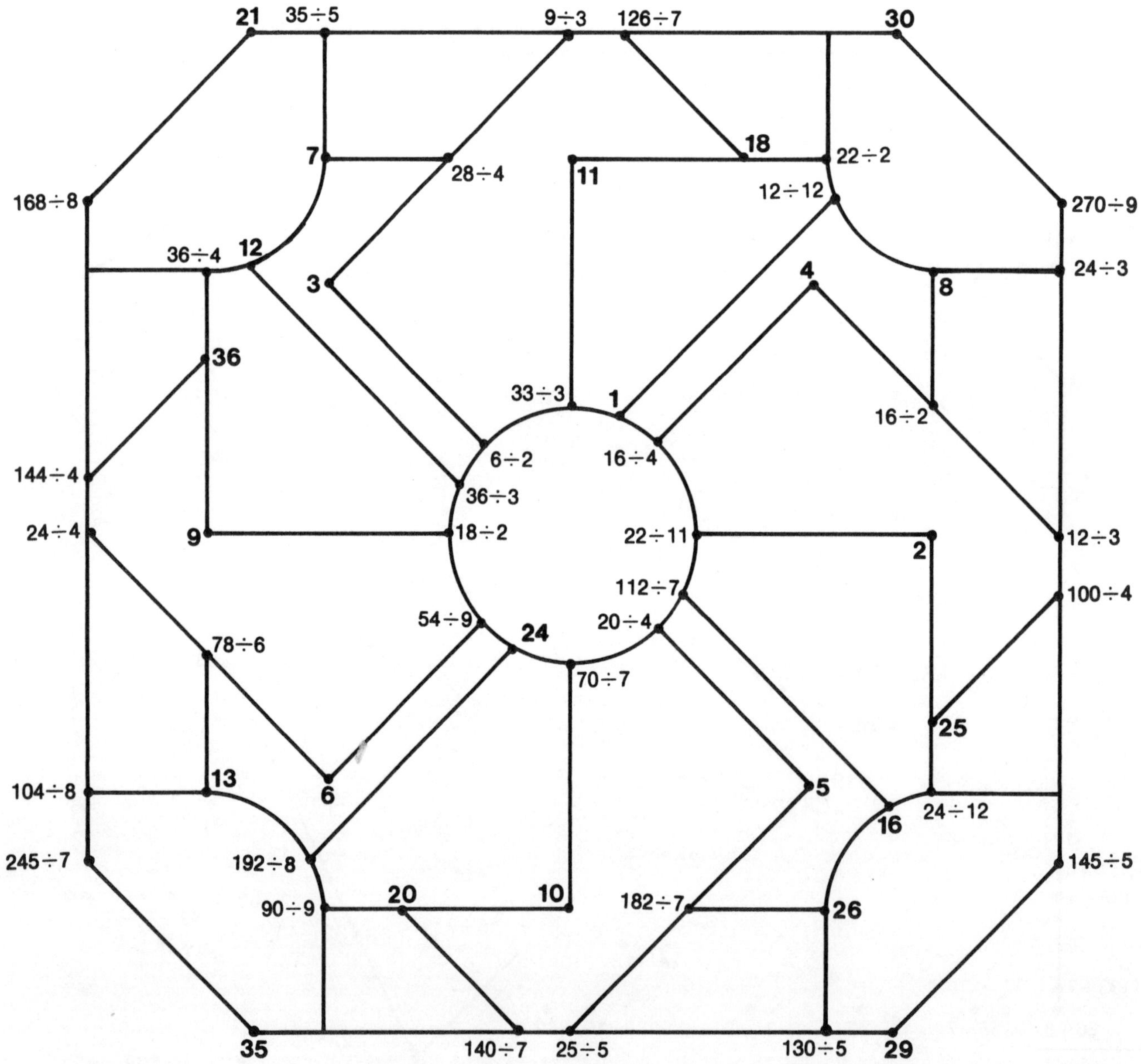

Division

Design 14

Name ______________________

Date ______________________

1. **Read all of the directions before you begin.**
2. **Complete one division problem next to a dot.** (Use scratch paper if you need to.)
3. **After you complete the problem, find the answer on the paper.**
4. **Use a rule (or straight edge) and draw a line to connect the dot by a problem to the dot by the correct answer.**
5. **Complete all of the problems in the same way.**
6. **Make sure you do all of the problems. Some lines pass through dots that are not part of the problem and some answers are used twice.**
7. **Carefully color your design after completing all of the problems.**

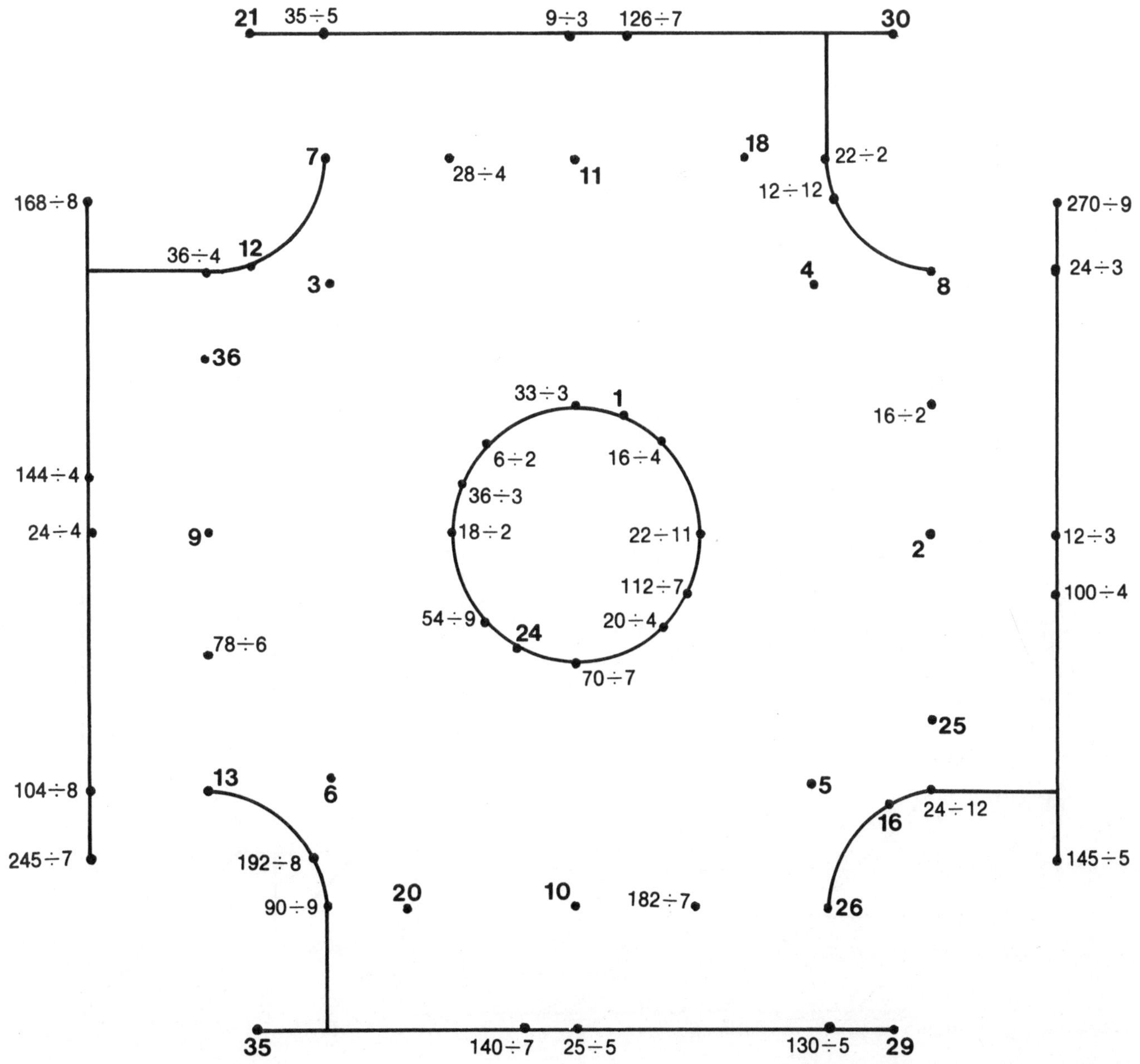

Division

Correction Key
Design 15

1. Use this design as a correction key.
2. Allow your students to correct their own work.
   Make a transparency of this design and instruct your students to place the transparency over their completed design for a quick and easy check.

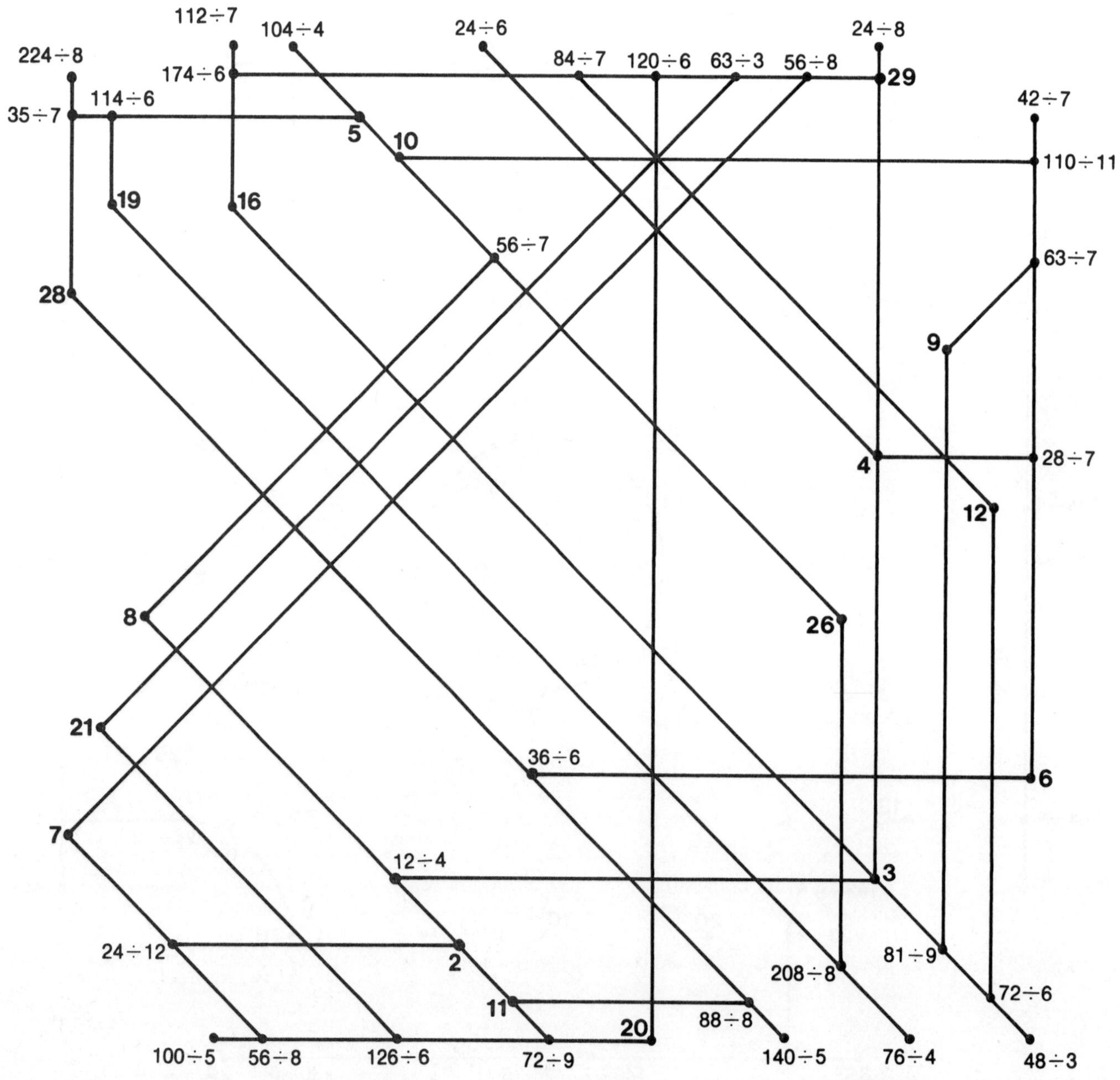

Division
Design 15

Name ____________________

Date ____________________

1. Read all of the directions before you begin.
2. **Complete one division problem next to a dot.** (Use scratch paper if you need to.)
3. **After you complete the problem, find the answer on the paper.**
4. **Use a rule (or straight edge) and draw a line to connect the dot by the problem to the dot by the correct answer.**
5. **Complete all of the problems in the same way.**
6. **Make sure you do all of the problems. Some lines pass through dots that are not part of the problem and some answers are used twice.**
7. **Carefully color your design after completing all of the problems.**

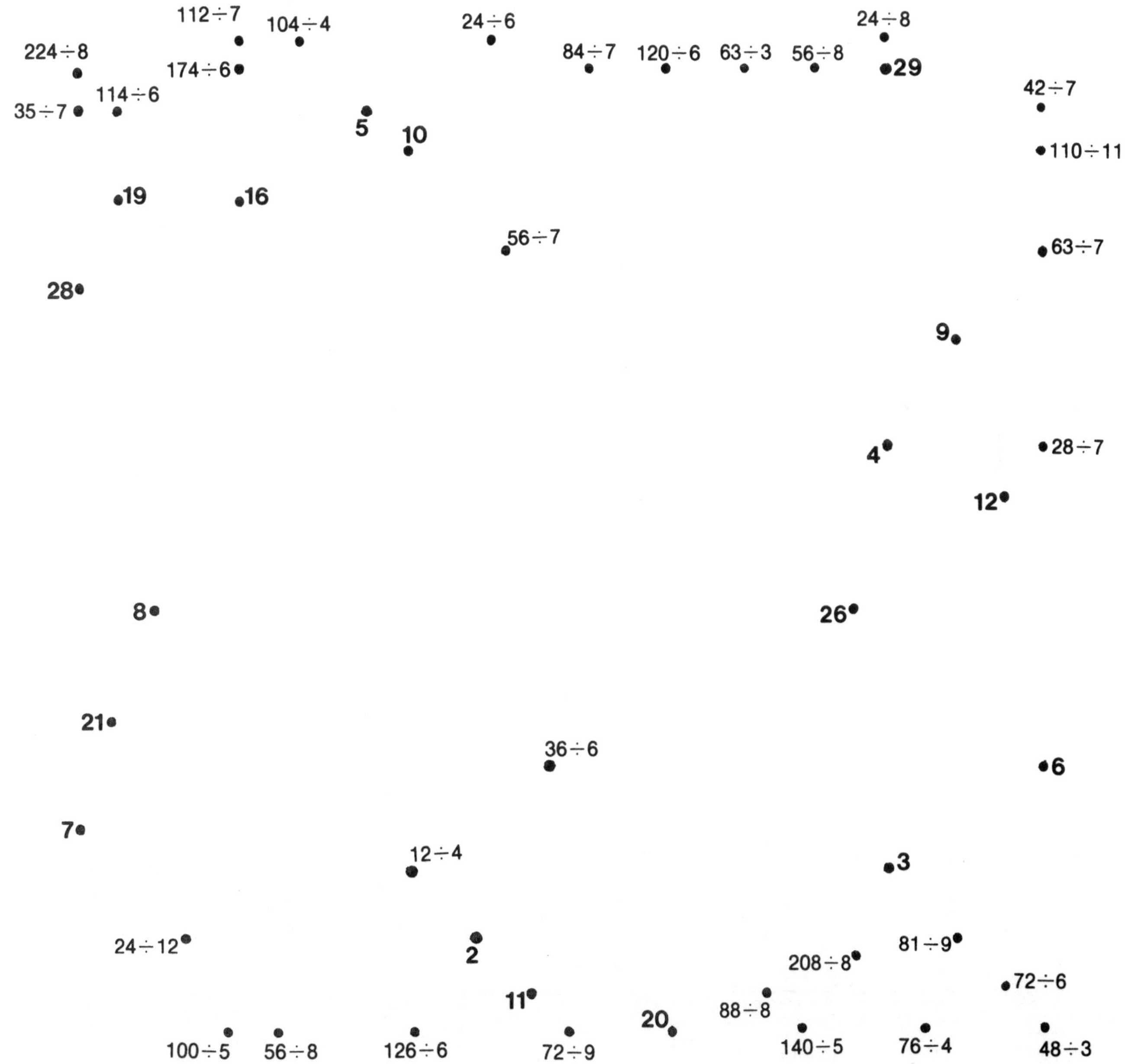

Division

Correction Key
Design 16

1. Use this design as a correction key.
2. Allow your students to correct their own work. Make a transparency of this design and instruct your students to place the transparency over their completed design for a quick and easy check.

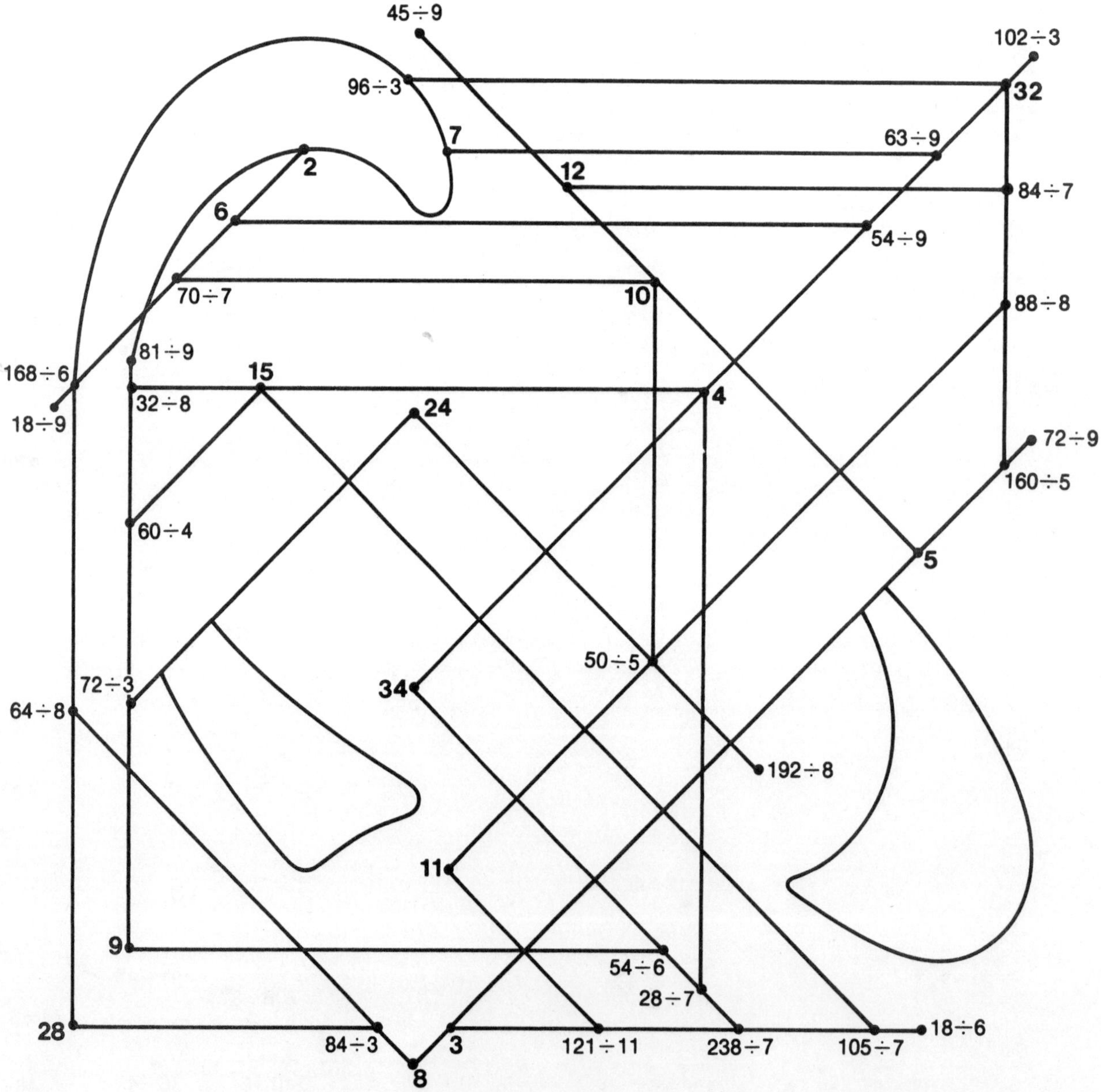

Division
Design 16

Name ____________________

Date ____________________

1. Read all of the directions before you begin.
2. Complete one division problem next to a dot. (Use scratch paper if you need to.)
3. After you complete the problem, find the answer on the paper.
4. Use a rule (or straight edge) and draw a line to connect the dot by a problem to the dot by the correct answer.
5. Complete all of the problems in the same way.
6. Make sure you do all of the problems. Some lines pass through dots that are not part of the problem and some answers are used twice.
7. Carefully color your design after completing all of the problems.

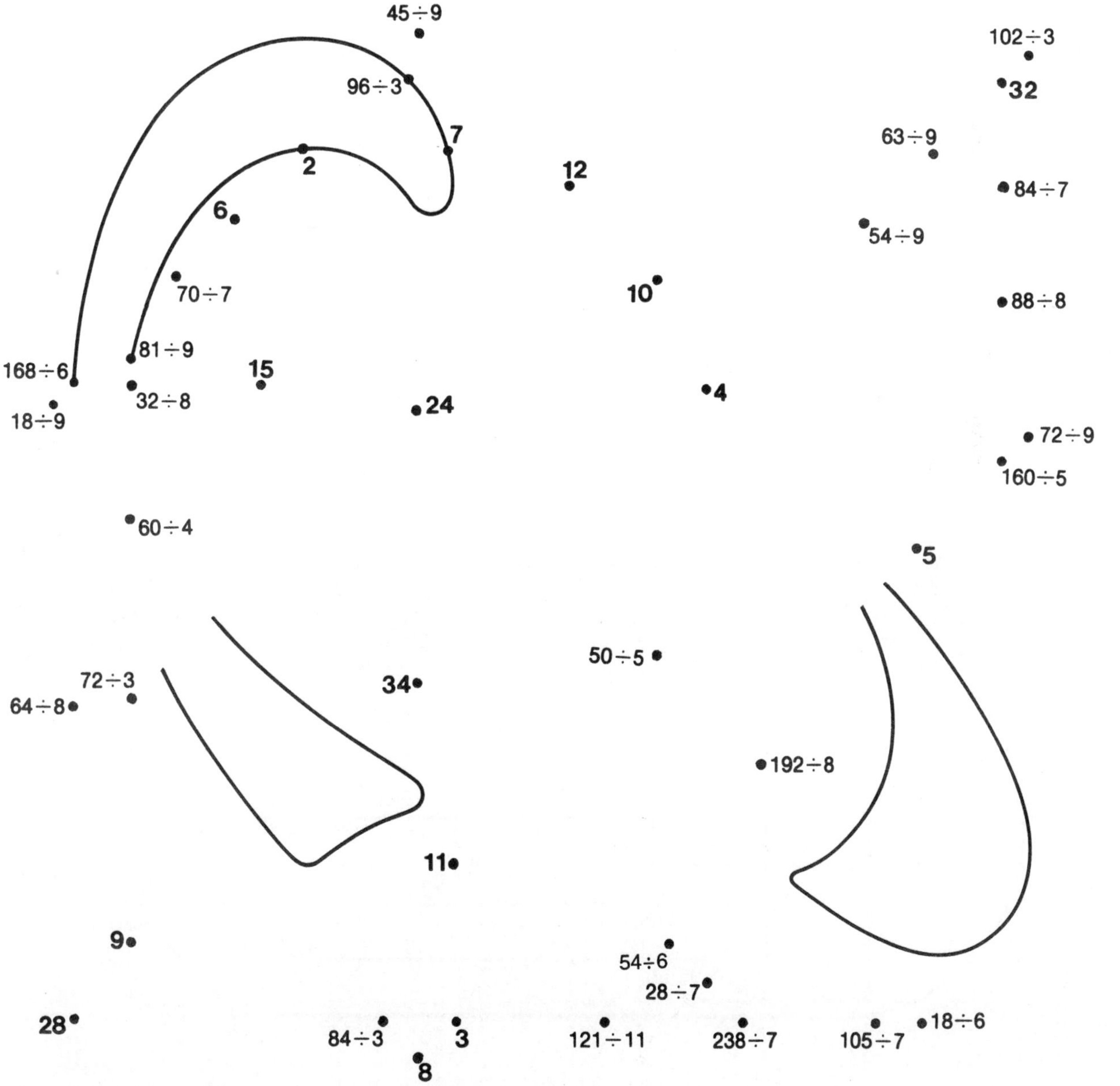

Division

Correction Key
Design 17

1. Use this design as a correction key.
2. Allow your students to correct their own work.
   Make a transparency of this design and instruct your students to place the transparency over their completed design for a quick and easy check.

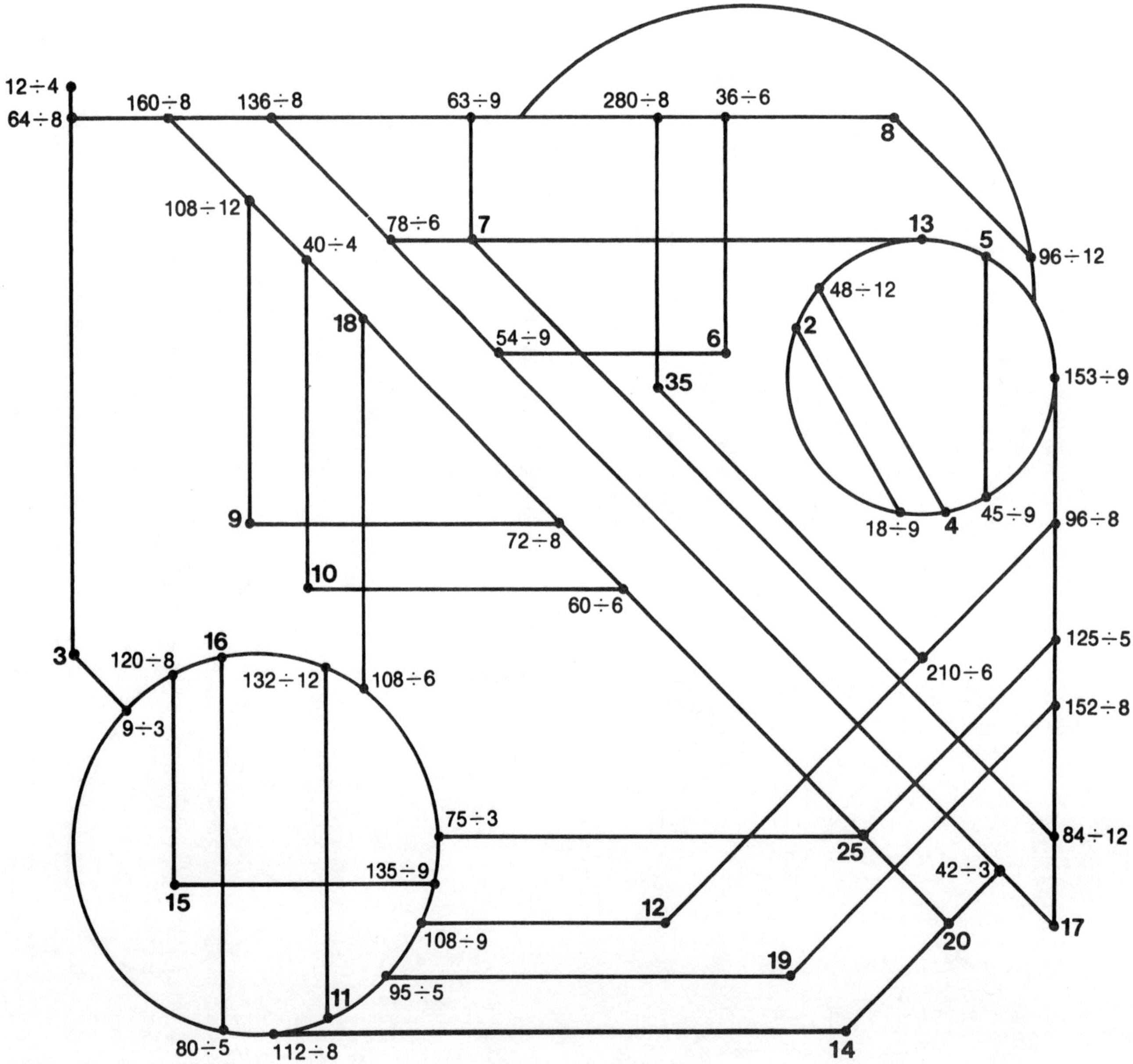

Division
Design 17

Name ______________________

Date ______________________

1. Read all of the directions before you begin.
2. Complete one division problem next to a dot. (Use scratch paper if you need to.)
3. After you complete the problem, find the answer on the paper.
4. Use a rule (or straight edge) and draw a line to connect the dot by a problem to the dot by the correct answer.
5. Complete all of the problems in the same way.
6. Make sure you do all of the problems. Some lines pass through dots that are not part of the problem and some answers are used twice.
7. Carefully color your design after completing all of the problems.

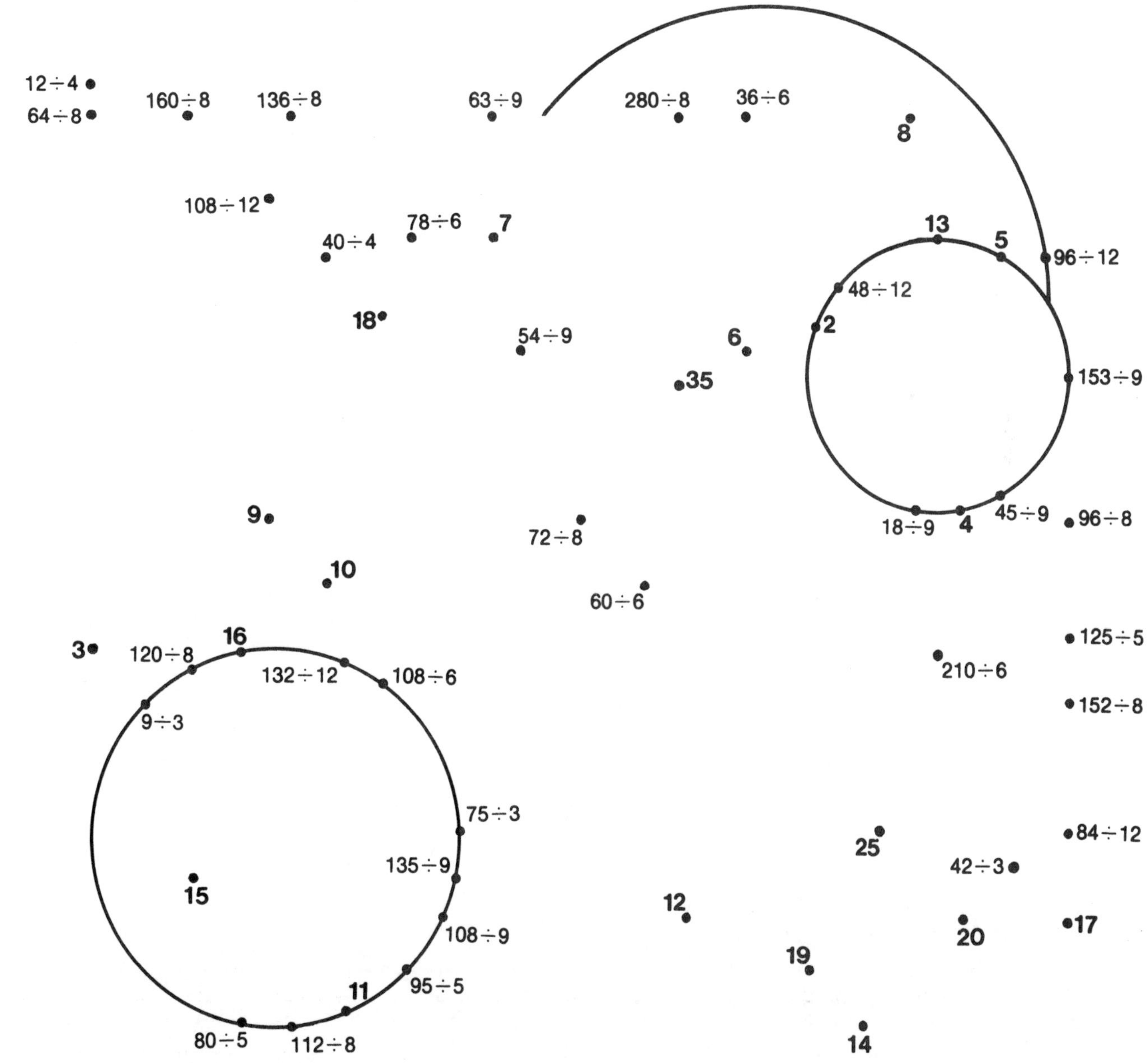

Division

Correction Key
Design 18

1. Use this design as a correction key.
2. Allow your students to correct their own work. Make a transparency of this design and instruct your students to place the transparency over their completed design for a quick and easy check.

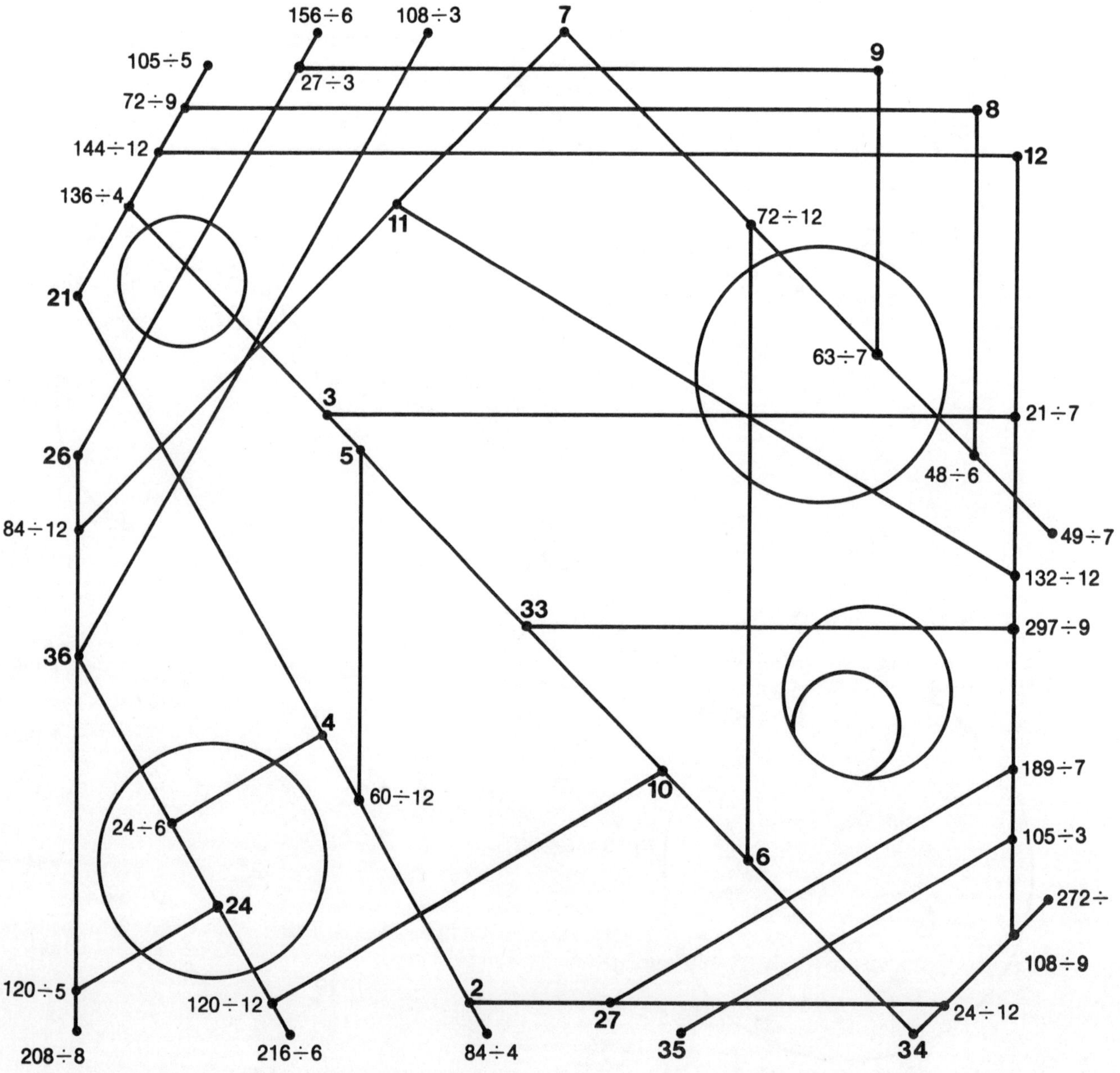

Division
Design 18

Name ______________________________

Date ______________________________

1. **Read all of the directions before you begin.**
2. **Complete one division problem next to a dot.** (Use scratch paper if you need to.)
3. **After you complete the problem, find the answer on the paper.**
4. **Use a rule (or straight edge) and draw a line to connect the dot by a problem to the dot by the correct answer.**
5. **Complete all of the problems in the same way.**
6. **Make sure you do all of the problems. Some lines pass through dots that are not part of the problem and some answers are used twice.**
7. **Carefully color your design after completing all of the problems.**

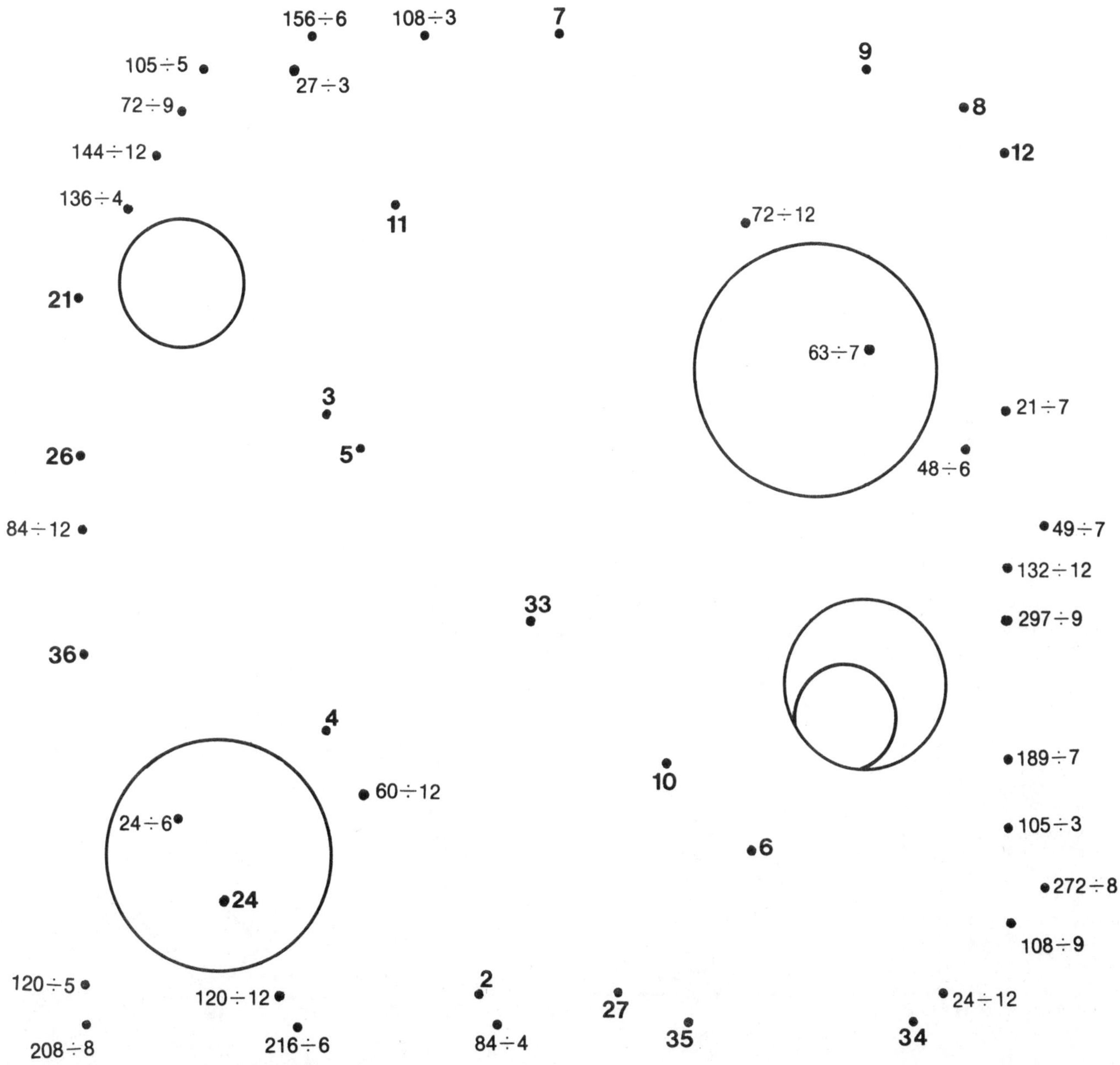

Division

Correction Key
Design 19

1. Use this design as a correction key.
2. Allow your students to correct their own work. Make a transparency of this design and instruct your students to place the transparency over their completed design for a quick and easy check.

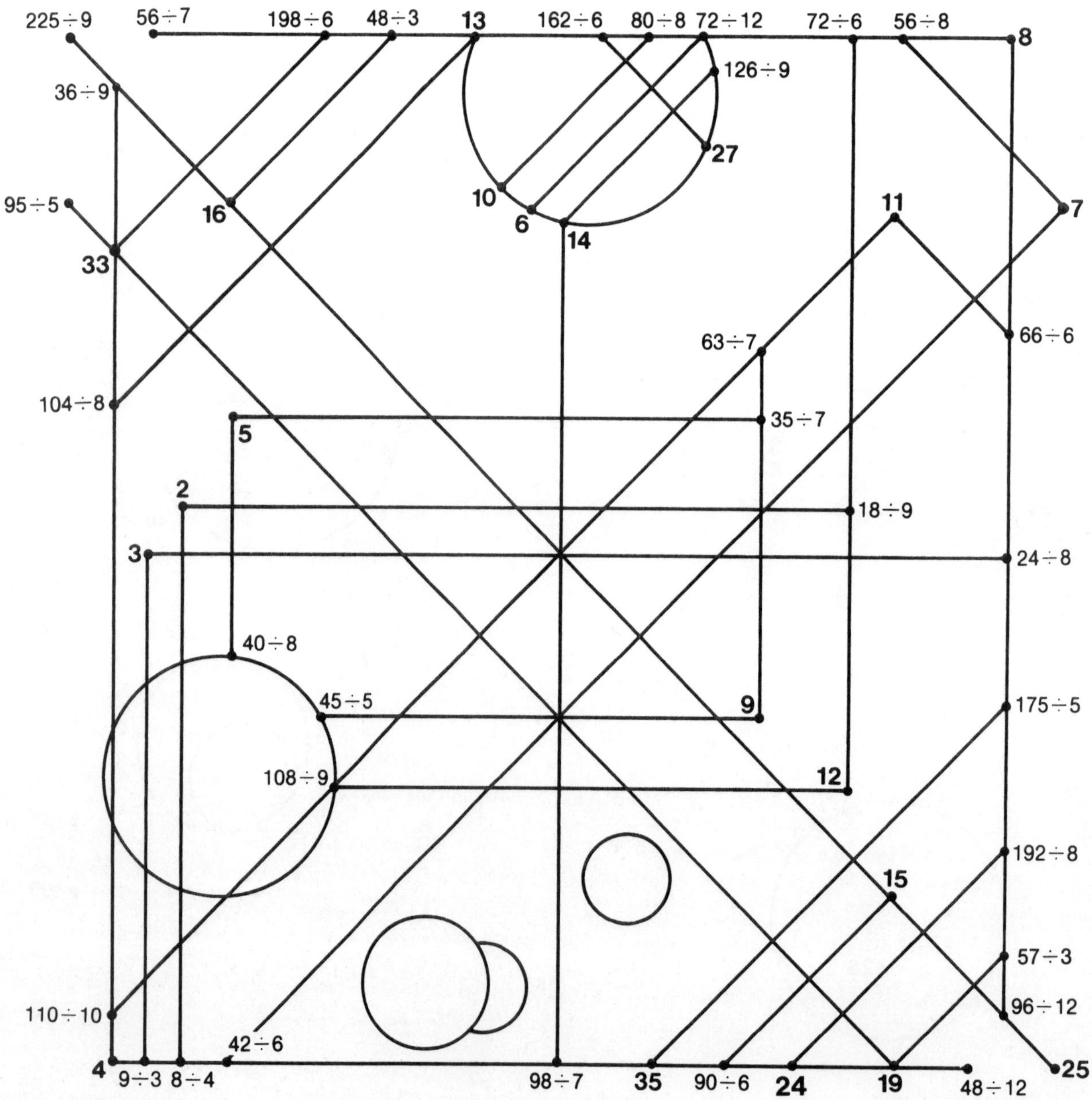

Division
Design 19

Name ______________________

Date ______________________

1. **Read all of the directions before you begin.**
2. **Complete one division problem next to a dot.** (Use scratch paper if you need to.)
3. **After you complete the problem, find the answer on the paper.**
4. **Use a rule (or straight edge) and draw a line to connect the dot by a problem to the dot by the correct answer.**
5. **Complete all of the problems in the same way.**
6. **Make sure you do all of the problems. Some lines pass through dots that are not part of the problem and some answers are used twice.**
7. **Carefully color your design after completing all of the problems.**

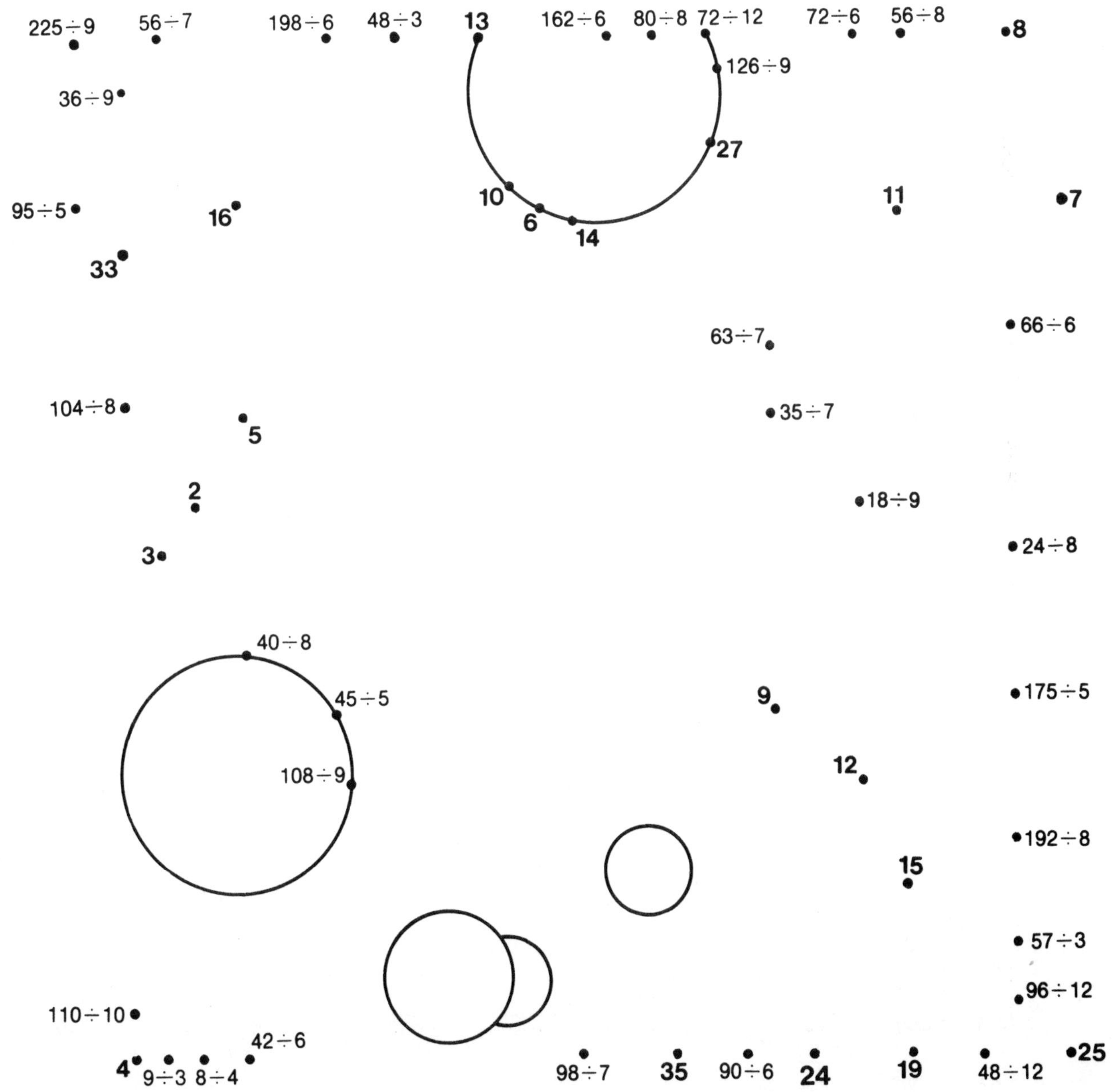

Division

Correction Key
Design 20

1. Use this design as a correction key.
2. Allow your students to correct their own work.
   Make a transparency of this design and instruct your students to place the transparency over their completed design for a quick and easy check.

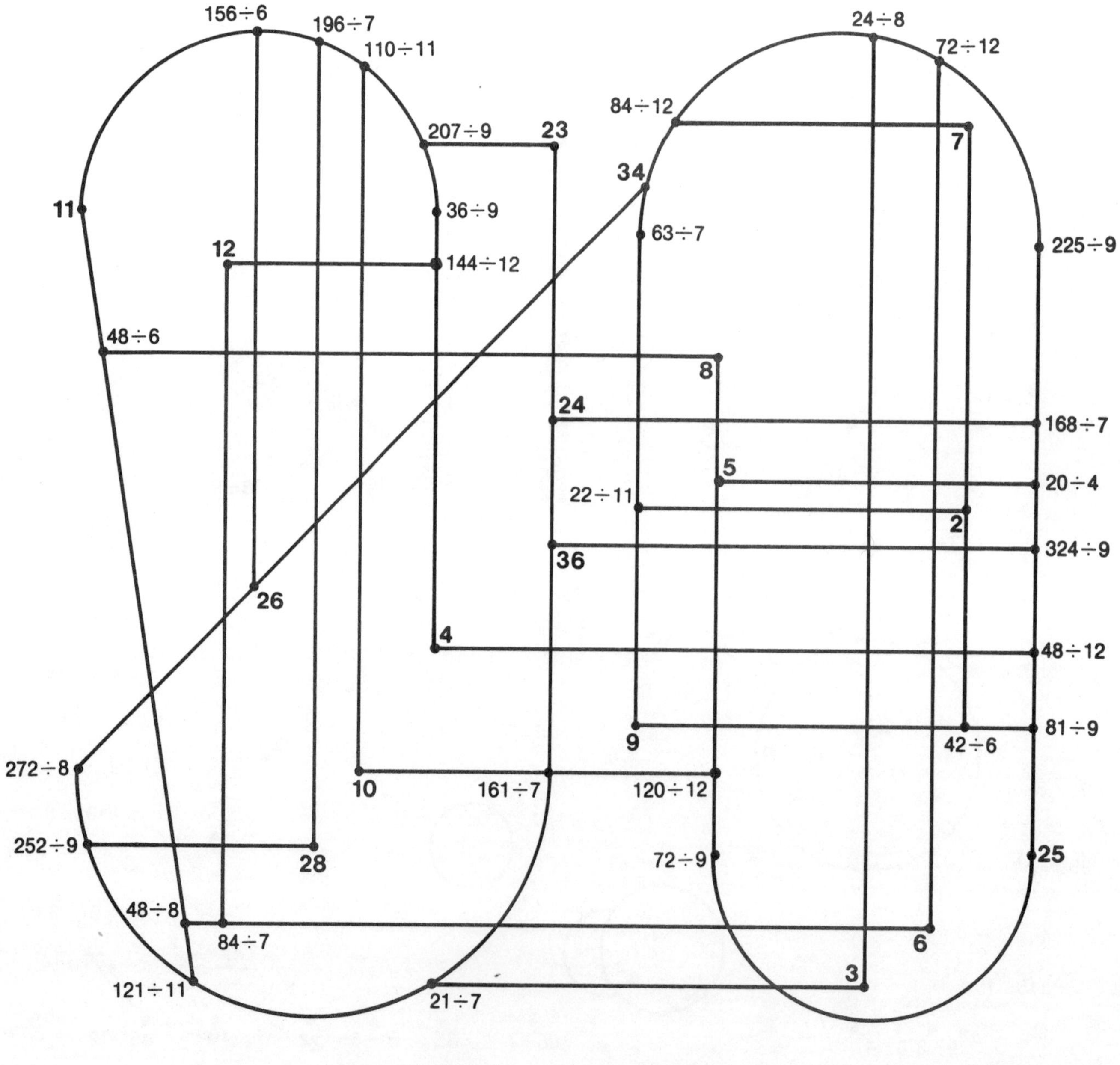

Division
Design 20

Name ______________________

Date ______________________

1. Read all of the directions before you begin.
2. Complete one division problem next to a dot. (Use scratch paper if you need to.)
3. After you complete the problem, find the answer on the paper.
4. Use a rule (or straight edge) and draw a line to connect the dot by a problem to the dot by the correct answer.
5. Complete all of the problems in the same way.
6. Make sure you do all of the problems. Some lines pass through dots that are not part of the problem and some answers are used twice.
7. Carefully color your design after completing all of the problems.

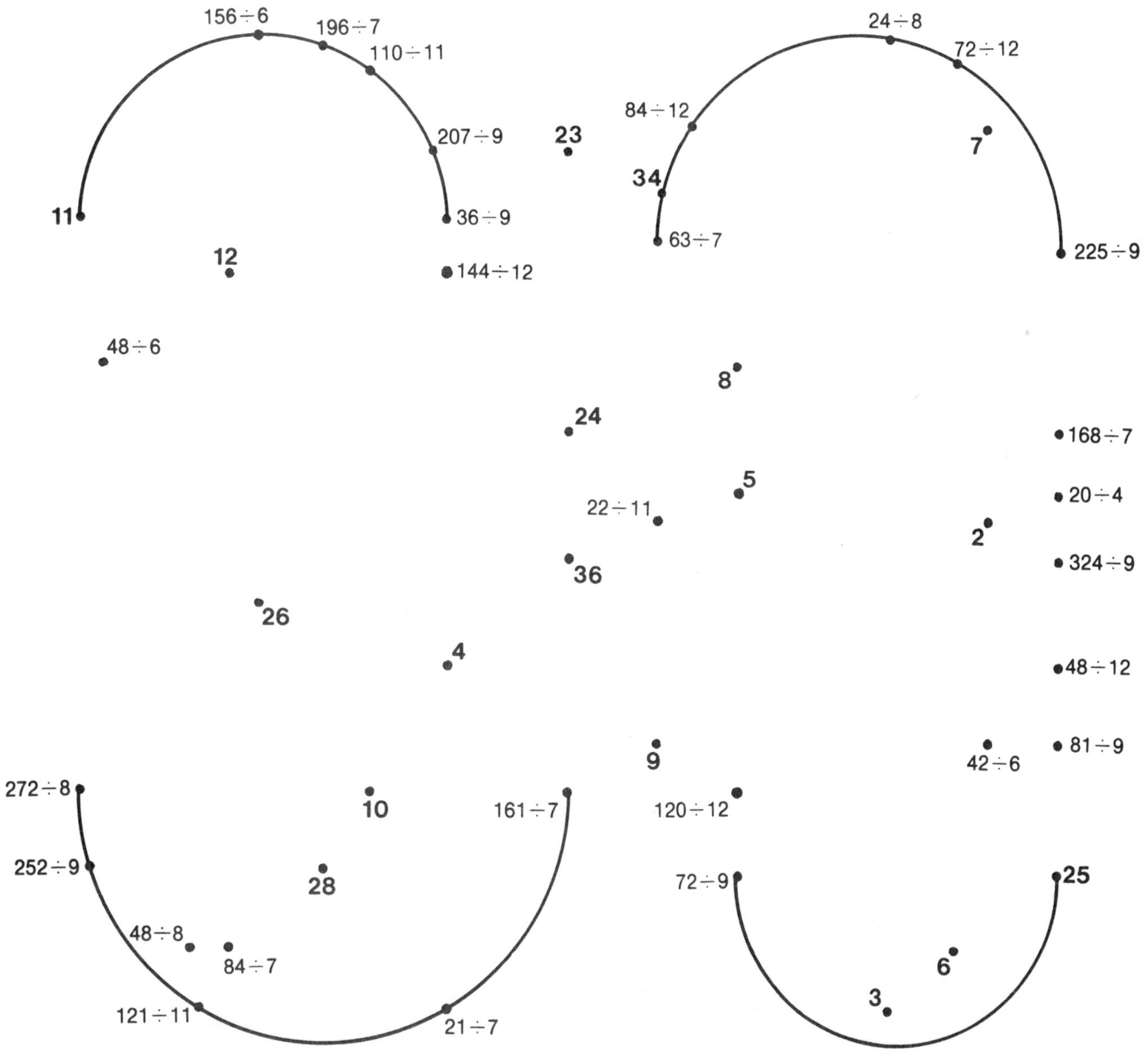

Golden Educational Center
G.E.C. Publications
"Leading The Way In Creative Educational Materials" ™

Designs in Math

Students create geometric designs by connecting the dots with a ruler between a math problem and the correct answer. These activities reinforce basic math fact memorization, following directions, fine muscle control, visual-motor skills, and principles of design. Each book contains 20 reproducible activities. There is a completed design that can be used for a correction key or made into a transparency for classroom instruction.

Addition #1006 .... 3rd-5th
Subtraction #1007 .... 3rd-5th
Multiplication #1008 .... 3rd-6th
Division #1009 .... 4th-7th
Fractions #1010 .... 5th-9th
Frac.-Dec. Equivalents #1011 .... 5th-9th

❖ ❖ ❖

Beginning Math Art

Similar to *Designs in Math,* these books show students the fundamentals of design while they practice solving their math problems and then connect the correct dots with a straight edge. There is a maximum of 12 problems per design. Young students will enjoy completing the 20 activity pages in each of these books. *Teacher instructions and answer keys are included.*

Add & Subtract 0-10 #1013 .... K-1st
Add & Subtract 11-20 #1014 .... 1st-2nd
Mult. & Divide 0-12 #1015 .... 2nd-3rd

❖ ❖ ❖

Multiplication Practice Puzzles

The activities in this book have students complete a multiplication problem and then a subtraction (or addition) problem using the answer to the multiplication problem. After determining the final answer, they color the shape containing the problems the correct color, which is determined by the color key on the activity page. Each picture has 30 to 50 problems per page. The book has 60 different puzzles (pictures) and answer keys.

Grades 3rd-7th #1025

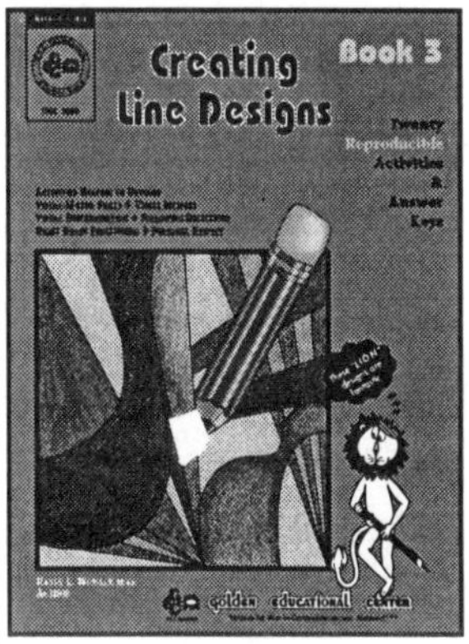

Creating Line Designs

These are a well-liked series of workbooks that incorporate using a pencil and ruler to connect points on a page. In the *CLD* books, a completed pattern is shown as an example for the child to duplicate. Each book is progressively more difficult in design, with 20 designs per book. Students are motivated to master more difficult designs and consequently to develop their visual perception and memory. We've also called them "pre-drafting skill workbooks," as they are designed to get students to master ruler and pencil skills.

Book 1 #1001 .... 3rd-5th
Book 2 #1002 .... 3rd-5th
Book 3 #1003 .... 3rd-6th
Book 4 #1004 .... 4th-7th

❖ ❖ ❖

Read+n+Draw
Following Directions

These great workbooks teach children to follow directions through exciting measuring/drawing activities. Students increase reading comprehension by reading sequential directions, plotting points and completing the Twenty activities in each book make learning fun. *Teacher instructions and keys are included.*

Book 1 #1021 .... 3rd-5th
Book 2 #1022 .... 4th-8th

❖ ❖ ❖

Using a Visual Grid for Solving Math Word Problems

Paring down to the most essential information, young students can now see and even understand the most essential elements/terms of a word problem. There are 30 lessons in each of these books. *Teacher instructions and answer keys are included.*

Add & Subtract 0-99 #2221
(no borrowing/carrying) K-2nd

Add & Subtract to 999 #2222
(with borrowing/carrying) 1st-3rd

Mult. & Division #2223 3rd-5th

Measurments, Time & Money
Mostly Mult. & Div.) 4th-7th

❖ ❖ ❖

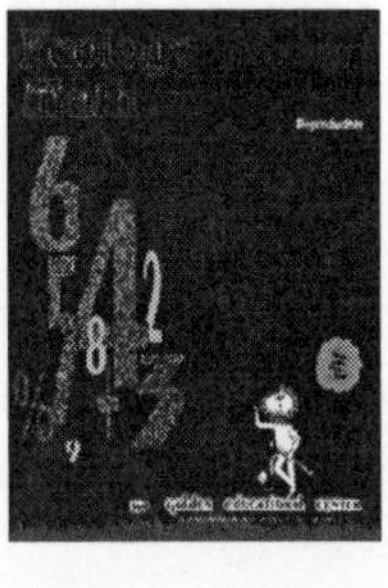

Ecology Math

Students are presented with simplified ecological information. They then write their ideas on how they can help the ecological situation. The lessons continue with mathematical problems referring to the information first given. There are 17 math lessons, as well as 17 discussion and information sheets. *Teacher instructions and answer keys are included.*

Grades 5th-8th #1105

golden educational center
G.E.C. Publications
"Leading The Way In Creative Educational Materials" ™

Country Study Books

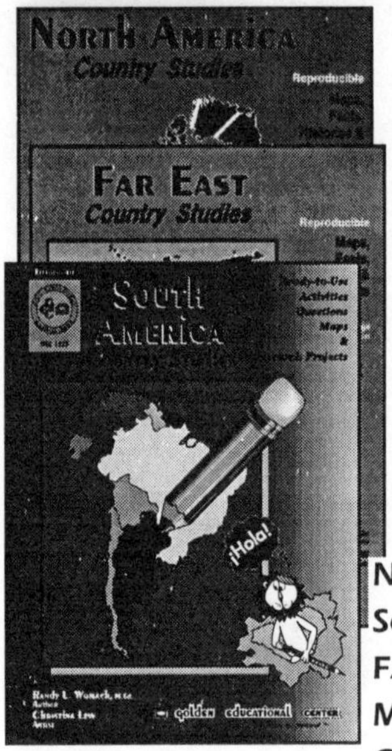

Each individual country within each respective continent has new words to look up in a dictionary, a large outline map, one page of current facts (population, area, capital, largest city, flag w/description, additional interesting information and more), one or two pages of history through independence and a page of review questions at the end of each section. Bonus (research) activities are also at the end of each section. *96 to 112 pages; Teacher instructions and keys are provided.*

North America #1965 .... 4th-8th
South America #1975........ 4th-8th
Far East #1935 .... 4th-8th
Middle East #1936........... 4th-8th
Canada #1985................ 4th-8th

❖ ❖ ❖

California ❖ Washington ❖ Texas

California Early History:
This is a simplified, yet complete, resource book detailing the history of California through statehood. This tremendous resource includes sections on the contributions made by Native Americans, Asians, Africans and Mexicans to the growth and development of the state. Review questions and bonus (research) activities follow each section. *94 pages; Teacher instructions and answer keys are provided.*

California Geography:
Students are given a world overview with a specific review of California's climate and its physical, economic and political features. Lessons are reinforced with maps, exercises, review questions and bonus (research) activities. *63 pages; Teacher instructions and answer keys are provided.*

Washington State Geography:
Identical to California, but with Washington's information. *63 pages; Teacher instructions and answer keys are provided.*

Texas Geography:
Identical to California, but with Texas information. *63 pages; Teacher instructions and answer keys are provided.*

California Missions & California Missions Poster:
Coming Summer 1998

California Early History #2911........ 4th-8th
California Geography #2912........... 4th-8th
California Missions #2913.............. 4th-8th
California Missions Poster #2914
Washington Geography #2106 .... 4th-8th
Texas Geography #2105 .... 4th-8th

Learning the Continents

Students use maps to identify, memorize and locate the countries, waterways and points of interest on each continent (Australia and Antarctica are not divided into countries). Each continent has 16 pages of map activities and a word search puzzle at the end for spelling review of the countries. *96 pages; Teacher instructions and answer keys are provided.*

#1906 .... 4th-8th

❖ ❖ ❖

Continent Maps & Studies

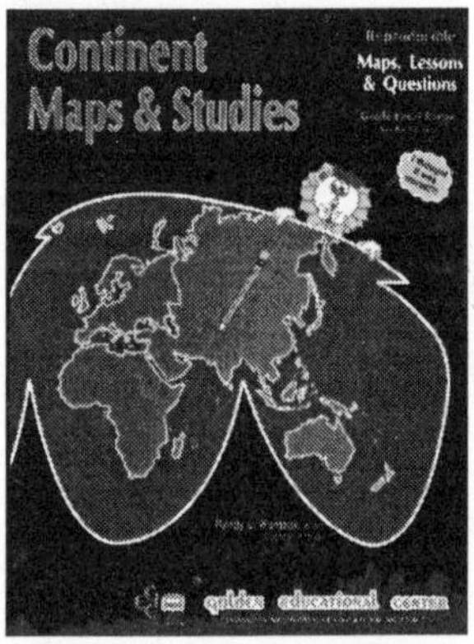

This book contains outline, waterway and political boundary maps and an individual fact sheet for each of the continents. (Antarctica only has an outline map and fact sheet.) There are questions, research activities and a glossary that can be used with each continent. *64 pages; Teacher instructions and keys.*

#1905 .... 4th-8th

❖ ❖ ❖

U.S. Studies

U.S. Outline Maps:
This book has an individual fact sheet and outline map for each state, Washington, D.C. and the entire United States. U.S. waterways and state boundary maps are also included. There are also individual question and research pages that can be used with each of the states. *112 pages.*

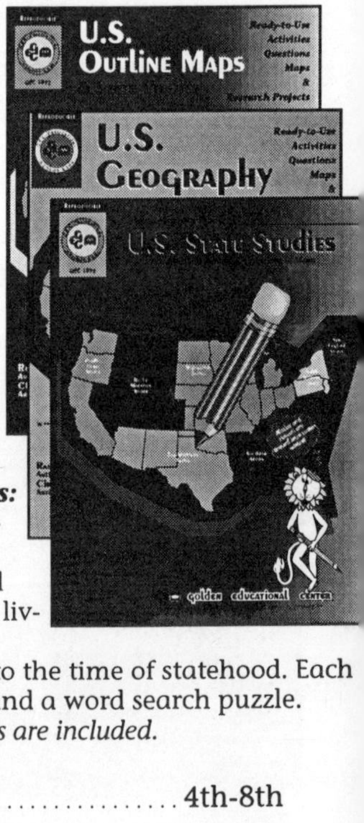

U.S. Geography:
Each section of this book begins with new vocabulary words to look up and define. The sections cover a world overview and physical, economic, political and climatic features of the country. Map activities, review questions and bonus (research) activities are included in each section. There is also a review section for the entire book with crossword puzzles, a word search and other activities. *80 pages; Teacher instructions and keys.*

Complete Book of U.S. State Studies:
This book includes the same information as in the *U.S. Outline Maps* book listed above. It also includes historical information about Native Americans living in the region before any settlers came, as well as the early settlers up to the time of statehood. Each state also has interesting trivia facts and a word search puzzle. *219 pages; Teacher instructions and keys are included.*

U.S. Outline Maps #1992 .... 4th-8th
U.S. Geography #1993 .... 4th-8th
Complete U.S. State Studies #1995........... 4th-8th